GEOEXPLORATION MONOGRAPHS

Series 1 – No. 8

GEOPUBLICATION ASSOCIATES

GEOEXPLORATION MONOGRAPHS

Series 1 – No. 8

Editors

N. A. ANSTEY, London

P. N. S. O'BRIEN, London

1976

GEBRÜDER BORNTRAEGER · BERLIN · STUTTGART

Seismic Reflection Interpretation

by

A. A. FITCH

with 56 figures in the text and on 1 folder

1976

GEBRÜDER BORNTRAEGER · BERLIN · STUTTGART

ISBN 3 443 13009 7

Printed in Germany by Konrad Triltsch, Graphischer Betrieb, Würzburg
Paper: Scheufelen, 7311 Oberlenningen
Type: Garamond

Contents

Editors' Preface

This monograph is meant for those who are concerned with interpreting seismic reflection records, especially those engaged in the search for oil and gas. Seismic interpretation is an art which needs to be based on a clear knowledge of a highly developed technology together with a proper understanding of what actually can happen within the Earth.

Many interpreters lack this understanding. But not A. FITCH who has had a world-wide experience in seismic interpretation and, after more years than he may care to remember, is still fully involved and at the forefront in the field of seismic prospecting. But more than that, he has on numerous occasions lectured to students and led symposia on interpretation procedures, and always with an enthusiastic reception from his audience. – He is a first class communicator as well as a first class interpreter and so we were delighted when he accepted our invitation to write this monograph.

However, as he would be the first to admit, interpretation cannot be learnt merely by reading a book. The learning process requires the constant comparison of one's predictions with the results of the drill. In this respect we are sure that familiarity with this monograph will significantly increase the interpreters' success ratio and, consequently, we are most pleased to present it to the readers of this Series.

P. N. S. O'BRIEN

N. A. ANSTEY

Introduction

This writing is at three levels: the three levels represent broadly the writer's spheres of communication.

Level 1 is thought to provide the reader with enough to make a useful contribution to any seismic interpretation project. It represents the substance of lectures and informal talks to graduate students in geology and physics in industrial and in academic groupings.

Level 2 is the substance of talks with professional geologists and geophysicists.

Level 3 is the substance of talks with research groups.

There are disadvantages in this. The logical development from one level to another is rarely presented. To include it would have lengthened the time of writing beyond that available, and would have taken the reader far outside the field of interpretation. A second disadvantage is that during writing and printing the pace of development is such that topics will move from one level to another. Research topics will be reduced to practice or discarded, for example.

The method adopted, where possible, has been to write in general terms, and to illustrate and describe specific cases. This involves some repetition, but experience suggests that it helps the presentation.

I wish all seismic interpreters success in their efforts to peer into the subsurface, our closest remaining unexplored part of the universe, and in some ways the most intractable for the explorer. I extend these good wishes not only to workers in the field of hydrocarbons, but to miners and civil engineers who, increasingly, turn to seismic observation for help in subsurface problems.

Acknowledgements

Acknowledgement is due first to Seismograph Service Corporation, who authorized the preparation and publication of this material while I was employed by Seismograph Service (England) Limited. I am pleased to acknowledge also the great help given by oil companies and others who have given permission for the publication of illustrative material: Shell U.K. Exploration and Production Limited, the Shell-B.P./N.N.O.C. Joint Venture, Petroleos Mexicanos, Seismograph Service Limited, and many others.

It gives special pleasure to acknowledge the help of those personally involved in the book preparation. The editors, Messrs. N. A. ANSTEY and P. N. S. O'BRIEN contributed much in the way of new ideas and corrections to old ones in a stimulating correspondence: the remaining errors and ommissions are mine. Messrs. P. N. S. O'BRIEN and P. E. F. HAWKINS have written sections which are acknowledged in the text. Messrs. P. E. F. HAWKINS, W. E. LERWILL and T. P. HUBBARD have helped with the final checking of manuscript during my absence from Europe. Mr. G. HARBRON has prepared many of the figures. Mrs. J. H. PLAISTER typed the manuscript and prepared many of the drawings: a labour of vast proportions for which I give my special thanks.

November 1974

A. A. FITCH
Quito, Ecuador

Since these acknowledgements were written I havc been saddened to learn of the death of P. E. F. HAWKINS, my friend and colleague of many years. His last work was the checking of this manuscript.

LEVEL 1

1 Some preliminary material

1.1 Introduction

In this chapter the geological significance of a seismic reflection is examined, and the fundamental seismic process which leads to the production of a reflection seismic section is reviewed. These two subjects, along with a knowledge of the structural and sedimentary aspects of geology, are the basic material on which the interpretation skills are built. Both subjects are treated from the point of view of the interpreter.

The seismic sections which are the subject matter of this writing on interpretation are plots of the observed time-series (suitably modified by processing), which are recorded in the seismic process. In its essentials this involves putting a seismic signal into the ground, and recording the output at regular intervals. The process is repeated along a line of traverse, so that the ultimate output is a succession of time-series, referred to regularly spaced observation points along the line of traverse.

These output traces are plotted to a suitable scale, usually in a variable area presentation. Such a presentation is the 'seismic section' which the following pages set out to interpret.

This bare outline of the seismic process is not adequate for those unfamiliar with the seismic reflection method. These readers are referred to Part 3 of *Anstey* 'The seismic reflection technique' (see Bibliography).

1.2 The 'obvious' character of seismic reflection sections

Most seismic sections communicate readily the essential features of the geology of the true earth section which they represent. On a reconnaissance level it is correct to say that the interpretation is obvious.

There is a similarity between reflections on a seismic section; bedding planes on a cliff section; and correlation lines on a subsurface section. The geologist treats these three classes of observation in the same way. They contribute to his regional geological concepts. Studies of possible source rocks for petroleum, fluid migration and hydrocarbon reservoirs can be based on any of these classes of observation.

Monoclines, major folds and faults, profound unconformities, can be recognized at a glance on seismic sections. For work of this kind small scale sections (5 cm=1 s and 5 cm=4 km or smaller), are most useful, since a large region can be brought into the field of view, and details and difficulties are less prominent and are easily forgotten. In such a reconnaissance the structure is seen in terms of elements which are kilometres across.

Exploration for oil, which commissions most seismic interpretation, is concerned with the emplacement of a drill bit whose size is measured in centimetres. The explorationist would like much more than the broad reconnaissance results of the obvious features. He would like to know, qualitatively, the geological significance of small and obscure features as well as the obvious ones: and he would like the features of particular interest accurately presented in quantitative terms. He would like to have the detailed and exact observations of the field geologist in a well-exposed area, or of a subsurface geologist working with closely spaced cored holes. This cannot be achieved: but the object of the interpreter is to advance as far as he can towards this objective.

The meeting of the European Association of Exploration Geophysicists in June 1973 included a round-table discussion. Its theme was to set out what exploration management of oil companies expected from the geophysicists. It was quantified by H. R. WARMAN, Exploration Manager for British Petroleum, who stated the need to be able to resolve a stratigraphic unit of 10 metres thickness.

It is, perhaps, the contrariness of nature which puts many of the interpretation difficulties in the immediate neighbourhood of potential oil traps. For example, short, steep dip reversals; faults; reefs; pinchouts can all produce local complications for the interpreter.

Thus the work of the interpreter is often 'obvious' and straightforward in areas of little or no commercial interest for petroleum, difficult and complicated in the areas of possible accumulation. It is for this reason that a descriptive text must give much attention to the difficult parts of cross-sections which make up about 10% of the whole.

1.3 The assumptions made

The fundamental assumption made in seismic interpretation is that there is a one to one correspondence between a seismic arrival – as presented on a seismic section – and a geological horizon.

Experience shows that this is broadly true; but even at an elementary level the definitions must be sharpened a little.

Geologists define stratigraphic levels in a number of ways. The basic definition is in terms of time. A geological horizon is defined as the top surface of the sediments laid down at a given moment in time – say 100 million years ago. It can be defined with a time-stratigraphic or chronostratigraphic name, such as the top of the Mesozoic, which may not be defined with precision in terms of years. Such a time-stratigraphic name is applied by considering geological criteria for contemporaneity of deposition, by working outwards from some type locality at which an age was first defined. In this way the 'Permian' has been extended outwards from the original definition in Perm, a Russian locality, so that the Permian can now be recognized over most of the world.

Thus the definition of a geological horizon rests on a direct age determination in years, or on the 'contemporaneity' tests.

In practice, neither of these tests can be applied at points sufficiently close together to serve as the regular control for surveys of a detailed character, whether surface or subsurface surveys. Instead, characters which are more readily observed, with a greater density of observation, are substituted. Neither coincides with the time based definition of an horizon: but over the limited area of a seismic survey the agreement is usually close enough for the substitution to be made.

A first alternative is to use a 'bio-stratigraphic' subdivision. This is based on the appearance or disappearance of some key fossil, or 'zone' fossil, which is used to define the zone by its presence. A typical example of such a definition is the top of the zone of Stephanoceras humphriesianum, a Bajocian (Middle Jurassic) horizon. A bio-stratigraphic boundary will differ slightly from a time-stratigraphic boundary because animals and plants can migrate to new areas or disappear from the old under environmental pressures, and so their disappearance or appearance is not everywhere strictly contemporaneous.

Sometimes, where microfauna or microflora are used to determine stratigraphic units, statistical measures of the population will be used, rather than appearance and disappearance of a single zone fossil. This is subject to the same weaknesses as the use of zone fossils.

A second alternative is the use of a litho-stratigraphic division. This divides the geological sequence into the mapping units of the surface geologist or the subsurface formations of the subsurface geologist. A good example is the top of the Chalk in the northern part of the North Sea. Seismic reflections arise at these litho-stratigraphic divisions.

Reflections are generated at interfaces which may be bedding planes, fault planes, intrusive contacts or any other boundary between rock types which is sufficiently extensive. The required condition is that there should be a change of acoustic impedance across the boundary. Acoustic impedance is the product of compressional wave velocity and density, and it will be seen that cases may arise where an increase in velocity and a decrease in density across a boundary may result in little change in the acoustic impedance, and so the boundary generates only a weak reflection. Statistically, however, the two properties tend to change in the same direction.

Some features, such as change of colour or of grain size may have no effect on velocity or density. Others, such as the occurrence of lignite in the lower part of the Wilcox sand (Eocene) of the Gulf Coast, may give rise locally to reflections, although the lithological feature is not regarded as of sufficient importance or regularity to be used as a unit boundary. Some other boundaries, such as that between a dry rock and a wet one, give rise to reflections. These are not 'lithological' in the correct sense. They are related, not to litho-stratigraphic features, but to features such as a water-table, or a gas-oil or gas-water interface.

Lithological units are less extensive, and depart further from time-stratigraphic horizons, than units defined bio-stratigraphically. For example, sands and muds may be deposited contemporaneously, but they are different lithological units.

For the purpose of this book the definition of seismic arrival of interest is a primary, compressional wave reflection (or diffraction).

1.4 The methods of detailed seismic interpretation

The work of seismic interpretation uses but two important steps.

Continuity is the property observed on a seismic section whereby the pulse shape which is the seismic arrival of a reflection can be recognized on successive traces with, perhaps, small changes in arrival time from trace to trace. These repeated pulses create an alignment, and this alignment has continuity which can be followed. A length of continuity represents an 'island of confidence' from which one can work in both directions. The visual impression is dominated by the alignment, not by the individual pulses.

Recognition of continuity requires a curious re-orientation of the attention. Throughout the stages of seismic data-gathering and data-processing one thinks exclusively of a time-series, the vertical direction on the cross-section. The interpreter, in his concern for continuity, views the nearly horizontal dimension. It is often easier to transfer the attention to the near horizontal as the trace-width is decreased: this accentuates the continuity and lessens the emphasis on the traces themselves and the pulse forms which they contain.

Correlation is purely pattern recognition. The pattern may be a single pulse, distinguished by its length, amplitude or shape. More often a sequence of reflections is considered. In addition to the characteristics of the individual reflections, the spacing between them and the order contribute to the definition of a pattern.

Correlation is used primarily to relate one 'area of confidence' to another. Continuity may be interrupted by poor data in which the continuity cannot be traced. It may be interrupted by geological phenomena such as faulting. It may be interrupted simply by a gap in the survey. Whatever the reason for the interruption, the gap can be bridged by correlation.

Correlation affords a running check on the tracing of continuity. If continuity has been traced through a group of horizons, then the beginning should correlate with the end. In the course of interpretation this check must be made frequently.

1.5 Continuity

Geologically, the elements whose continuity is considered differ from those used by the seismic survey. Typically, the geologist is concerned with formations or their subdivisions.

It is a very old geological principle – the Principle of Superposition – that the different beds follow in the same order in different places where they are observed. Nowadays we should restrict this by some such phrase as 'as far as the palaeogeographic conditions remain sensibly unchanged'. With this proviso, formations and other geological divisions have continuity.

Seismic continuity of a reflection is *not* an expression of the continuity of a geological unit. It is an expression of the continuity of two geological units, one following immediately on the other; and at their contact is the interface at which the reflection is produced. Continuity of a seismic horizon requires that both formations are continuous. If there are facies changes, they must be gradual in character, and they must be such that the contrast at the contact is retained.

Seismic continuity results from the recognition, by eye, of corresponding shapes or elements in the trace form of successive traces, at almost the same time, such that the eye recognizes an alignment. Presentation of the trace form in the present day is generally in the variable area mode. In this mode the axis of time for each trace is vertically downwards. The right-hand side of the trace is black, the left-hand side is white, and the boundary between is shaped by a scaled measure of the ordinate at each time-sample. In other words, the trace is a plot of the amplitude with time, blacked out on the right-hand side of the track. The appearance of the section is affected by the 'dead trace width', the position of zero amplitude within the track, and by various modifications of the simple variable area presentation. The illustration, fig. 1, shows examples of good, poor, and doubtful continuity. The eye is able to follow continuity on traces whose signal-to-noise ratio is considerably less than 1 : 1.

The limits to confident interpretation are set by these signal-to-noise ratios. Although it is possible to recognize alignments at low ratios, it is difficult to identify the reflections one is following, and to discriminate, for example, between these reflections and interference crossing at an angle.

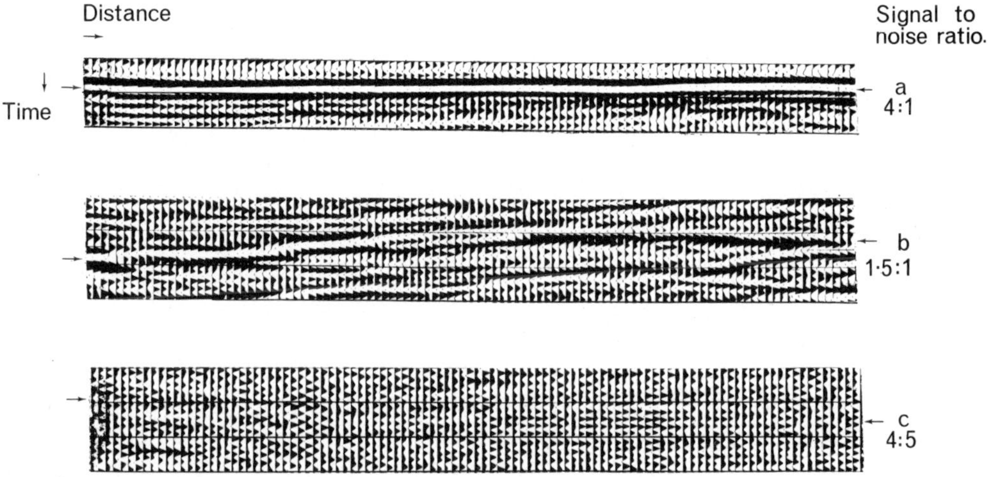

Fig. 1. Continuity.
(a) Good continuity, (b) Poor continuity, (c) Doubtful continuity.
To appreciate (b) and (c), look along the line of continuity with the eye just above the plane of the paper – and from either end.

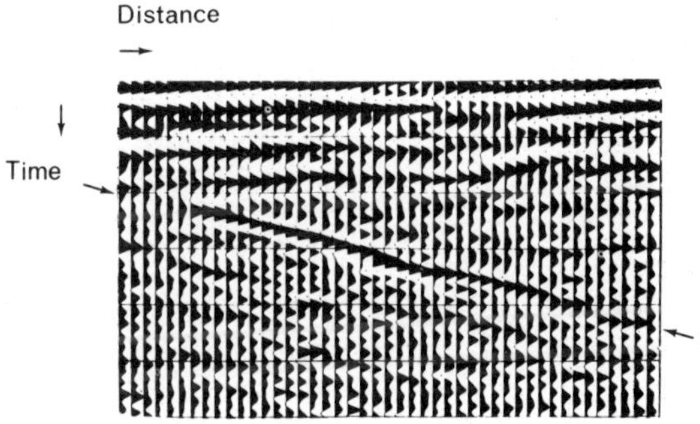

Fig. 2. Dipping continuity.
The impression of continuity is enhanced because the leading edges of the pulses (upper edges) are almost in a straight line. Horizontal axis represents distance, vertical axis represents time.

There is an interesting effect produced at a certain angle of dip. The individual pulse is almost triangular in shape: if the dip has the right value, then one side of each of these traingles is brought into a common alignment, and the impression of continuity is much accentuated. This is shown in fig. 2.

There is little more to be said of continuity as such: either it is recognized, or it is not. It is in the interpretation of noisy reflections that experience remains the guide to how far one can press on with doubtful continuity. It is re-assuring to report that the work of skilled interpreters of, say, 10 years ago, carried out on poor quality sections, is largely confirmed by current work on much better material.

1.6 Seismic, geological and mathematical correlation

Geological correlation consists of the recognition of a sample or an outcrop as belonging to the same unit as another sample or outcrop, without using the evidence of continuity from one location to the other.

The criteria used to establish correlation are quite varied, and they vary with the nature of the correlation which one wishes to establish. The three types of correlation correspond with the three definitions of stratigraphic level already described under (1.3).

Time-stratigraphic correlation is made on the basis of age determinations in years, yielding results whose differences can be attributed to experimental error. Sometimes tests of contemporaneity can lead to quite exact correlation on a time basis. Volcanic explosions can result in widespread deposition of tuffs, often in thin layers in varied sediments. A small-scale Recent example is the layer of ash running through the City of London which resulted from the burning of the whole city by Boudicca (Boadicea) in 61 A. D.

Bio-stratigraphic correlation is widely used by stratigraphers, particularly to achieve correlation over greater distances than can be spanned by litho-stratigraphic correlation. It depends on the identification of zone fossils, which are indicative of the zones named for these fossils. For some studies the statistics of the population are the basis for zoning and correlation. This is especially the case where microfauna or microflora provide the basis for zoning, though statistical studies of larger fossils – the distribution of numbers between the species, or variation in dimensions or other parameters of an evolving population are also used for zoning and correlation of the sedimentary rocks.

Litho-stratigraphic correlation can use any or all of the lithological features as a basis for correlation. These may be macrofeatures – such as colour, texture, and other petrological or mineralogical features which are readily observable by eye. Use may be made of microscopic features of texture or composition: such as the shapes of sand-grains or the calcite-dolomite relationship in a limestone. Extractive processes may be used to establish a basis for correlation. Examples are the study of heavy minerals in a sandstone, of the clay minerals in a clay, or a chemical analysis in an evaporite layer.

Litho-stratigraphic correlation makes use also of the order in which rock types are deposited as a guide to correlation.

Seismic correlation depends on two features of the reflections. The first is the shapes of the individual pulses which make up the sequence of reflections. The second is the sequence of the reflections and their spacing.

The range of shape of a reflection is not very wide – and it will be seen later that the changes may be due to effects which are not lithostratigraphic in significance. The sequence is a reliable basis for correlation: it is, in effect, a seismic statement of the principle of superposition. The spacing of reflections is of more doubtful reliability. Over a short distance the spacing remains sensibly unaltered. Over longer distances thickening and thinning, changes in seismic velocity, unconformities, and other features, tend to change the spacing of reflections. In spite of these drawbacks, seismic correlation is effective. Some early seismic surveys relied entirely on seismic correlation of isolated records, and not on continuity, to establish structure. Seismic correlation is illustrated in fig. 3.

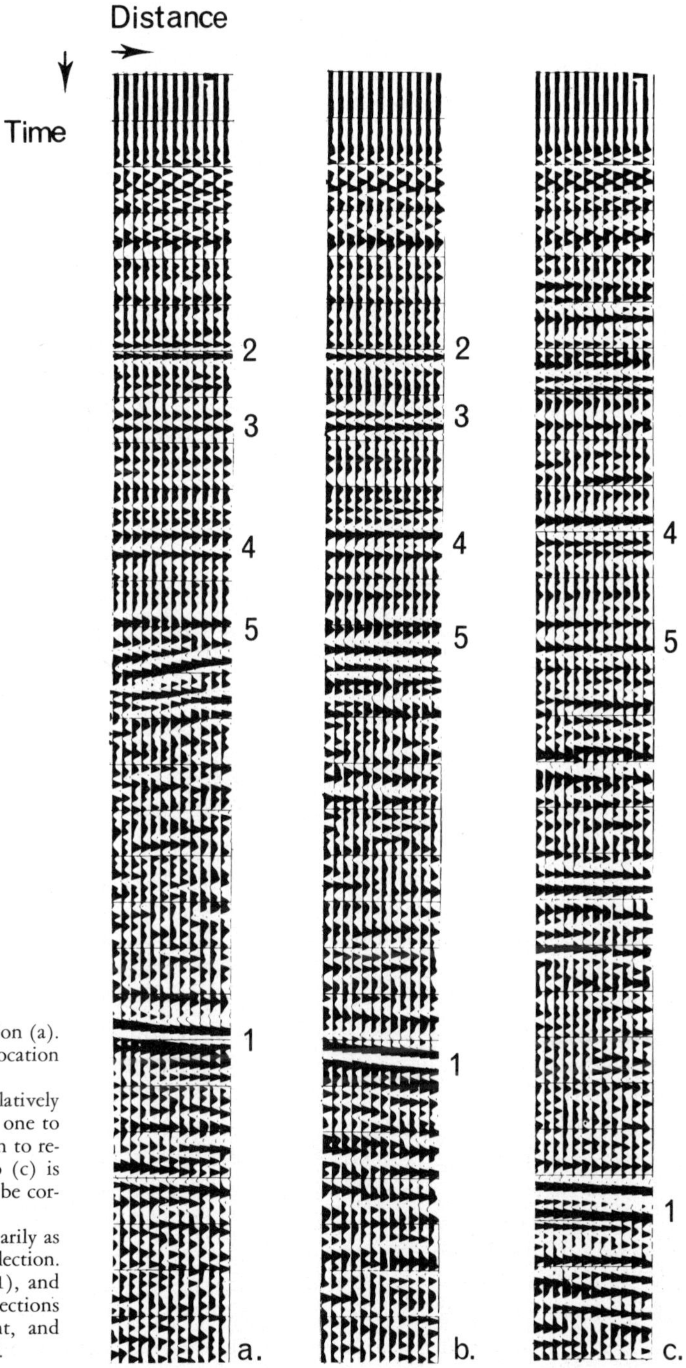

Fig. 3. Correlation.
Location (b) is 1 mile from location (a). Location (c) is 10 miles from location (a).
The step from (a) to (b) is relatively easy. Pattern recognition enables one to identify almost every arrival down to reflection 5. The step from (a) to (c) is less secure, and only 4 and 5 can be correlated with confidence.
Reflection (1) is recognized primarily as the one high amplitude, deep reflection. The interval between (5) and (1), and again below (1), contains reflections which are generally impersistent, and seismic correlation is not possible.

It is interesting, at this stage, to examine the mathematical process of correlation, and to establish the differences between this rigorous mathematical process and the geological and geophysical ones.

The mathematical prodecure is well described in standard texts. It is a manipulation of numbers, and so both geological or geophysical data must be presented as a regularly sampled series or as a continuous function. For geological data this can be achieved by logging in a well; or the measurement, in field or laboratory, of some property of the rock at regular intervals. For seismic data, the time-series which is the input to the cross-section plotter is the obvious choice.

The cross-correlation calculation is very sensitive to change in thickness or time-stretching of geological or geophysical data. Consider, for example, a typical seismic trace, 6 s long, with about 40 arrivals per second. If one introduces a thin intercalation of 12 or 13 ms length at 3 s, then this modified trace will produce a near-zero output when cross-correlated with the original. In thickness terms this is roughly a 20 m intercalation in the middle of a 10,000 m sequence.

The great merit of cross-correlation is that it is able to establish correlation in the presence of very high levels of noise. Noisy seismic data, or geological data subject to high measurement errors, can be correlated over short distances, and structural or time differences established, even when no features can be recognized in the data by inspection.

1.7 Tracing a seismic horizon

The primary purpose of most seismic surveys is to determine structure. This is achieved by tracing identifiable seismic horizons on cross-sections.

The routine is simply described. Where continuity can be recognized, it is lined in, using either a soft black pencil, or coloured peincils (coded to indicate particular horizons). Where continuity fails, a gap is left, and the lining in is resumed when the tests of correlation permit one to recognize the same horizons. Either peak or trough can be selected.

A practical note is that one should avoid the use of dark colours, heavily laid on, at least until the interpretation is complete and the uncertainties resolved. It is bad practice to preserve one's initial mistakes in interpretation in, say, bright red.

A much longer discussion will deal with what to do when the simple routine is not readily applied: but at this first level of treatment it can be seen that the seismic section is treated as an unusual kind of geological section. This is still adequate for the first appraisal of results.

At this point it may be useful to review the several features which contribute to mis-ties at the intersections of lines.

a) True changes in travel times. These can result from large movements of the water-table in land surveys, or from large tidal movements at sea.

b) Different assessments of stacking velocity in the two directions of survey, due to folding with its strike parallel to one of these directions, or some other cause. Use of different stacking velocities results in different reflection times.

c) Errors of survey. In heavy bush markers or reference points can disappear; and carrying the surveys for long distances from other control points can accumulate error. At sea occasional blunders arise as a result of losing the lane count in hyperbolic systems. More frequently the error is the normal error of the system. This is often quoted at the level of 67% probability. A little calculation will show that the mis-tie at intersection will often exceed the quoted error.

d) Differences from recording and processing changes, as described in Level 2 para 8.2, when two different surveys with different recording and processing procedures intersect.

e) Noise.

There is no rigid method of dealing with such mis-ties. What can be done is to raise or lower one survey so as to minimize the average mis-tie. Occasionally there is subsurface or sea-bottom evidence of a wrong survey position which can be corrected. In case (d) it is sometimes possible to calculate operators which transform the results with one system of recording and processing parameters to the equivalent with the other system and so reconcile results at intersection.

1.8 The splitting of reflections

In the pursuit of continuity or correlation, the interpreter will sometimes find that a reflection will split into two. Traced in the reverse direction, he will find a reflection which is terminated, and cannot be traced further. This is a troublesome phenomenon, because it is one which may have geological significance – or it may not. In this section the geological features which can cause such splitting are examined. The remarks bear on the tracing of reflections (1.7) and naming reflections (1.9).

I. Where a sequence is thickening, a single reflection will often split, and pass into two reflections. The process may be often repeated.

II. Where a sequence is changing: as for example from a shale overlying a sand into a series of shale-sand alternations, a simple reflection may give place to a sequence of reflections.

III. Overstep relationship at an unconformity gives rise to splitting of reflections.

IV. Overlap relationships, similarly, give rise to splits.

Where a seismic reflection splits, there is no strictly seismic reason for extending the continuity. Either the rock above, or that below, has changed, whichever direction has been traced at the split. This is the case where the split has geological significance. There is even less basis for taking one fork rather than the other if the split has a non-geological reason.

The continuity decisions are geological ones, and some generalizations can be made on the four cases described above.

(I) In a thickening sequence the detailed relationships can often be resolved into local overstep or overlap situations. The splits that result can be analysed as described below. In rare cases the increase in thickness appears to be quite uniform, without any evidence on the seismic section of a discontinuity. The best solution here is to continue the continuity in that element of the split which causes least departure from a smooth alignment. This is most likely to preserve the traced reflection at or close to the same stratigraphic level.

(II) Where facies changes lead to replacement of a single reflection by a number, usually of higher frequency, the same course is to be followed. Continue the continuity in that element that causes least departure from a smooth alignment.

In both these cases, tracing the reflection in the opposite direction leads to a similar rule. Where the reflection which is followed terminates, continue on the reflection element which least disturbs the smooth alignment.

(III) and (IV) Overstep and overlap conditions are illustrated in fig. 4. It will be seen that if the interpreter enters from the left he is faced with split reflections. In an overstep situation he goes up: in an overlap situation he goes down. The problem is to recognize the relationships in their seismic expression.

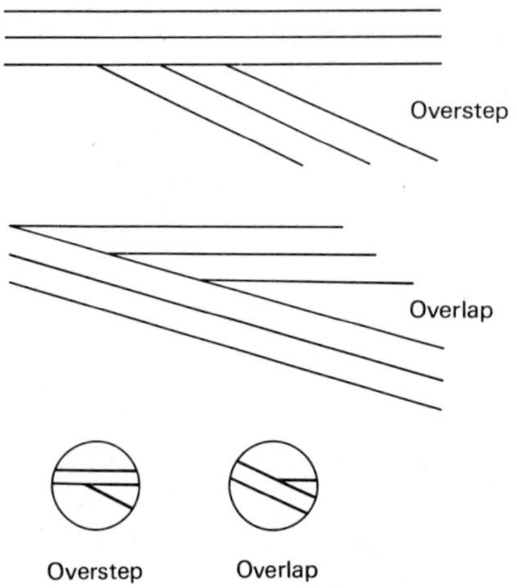

Fig. 4. Diagram of the geological relationships at overstep and overlap types of unconformity.
Circled are the corresponding 'splits' in reflections generated at these unconformities.
As so often in relating the real earth to seismic observations on it, one can reason confidently from the real earth model to its seismic response, less confidently from the seismic response to the real earth – which last is the task of the interpreter. In this case the uncertainty arises because there are other circumstances which will give rise to 'splits'.
It is largely the decision-taking at splits such as these which prevents complete automation of reflection picking in most areas.

The essential difference, evident from the figures, is that for the *overstep* configuration, it is the reflection immediately above the split horizon which is continuous and conformable with it. In the *overlap* configuration it is the reflection immediately below. If neither configuration can be recognized, the case is either (I) or (II). In the fig. 5 a is a case of overlap, 5 c is overstep, while 5 b and 6 are cases of thickening sequence.

The reverse direction of tracing offers no problem: the convergence of reflections leads the interpreter to the only solution.

These rules can be stated in a geological guise. In cases (I) and (II) there is no hiatus in sedimentation. There is a change in thickness in (I), and as will be discussed more fully later, this will give rise to a change in the seismic response which may be more obvious on the seismic section than the change in the geological section. In (II) there is a change of facies, which again induces a change in the seismic response. In both of these cases the rule requires the interpreter to follow the bedding plane as closely as possible through the thickening or changing sequence.

In cases (III) and (IV) there is a hiatus in sedimentation, and the reflection which is followed is generated by sediments before and after the hiatus. The rule requires the interpreter to follow this hiatus in sedimentation in each case, and it will enable him to do so, until the hiatus may disappear, and the continuity will have lead him to the correct place in the unbroken sequence. The essential difference between cases (III) and (IV) is that in case (III) the hiatus in sedimentation is filled in progressively from below, while in (IV) it is filled in progressively from above. Case (IV) involves an advancing shore-line. Case (III) involves an erosion interval.

It must be stressed that the discussion in this section is applicable where the 'split' or termination of the reflection results from geological causes. There are other reasons for such features: and

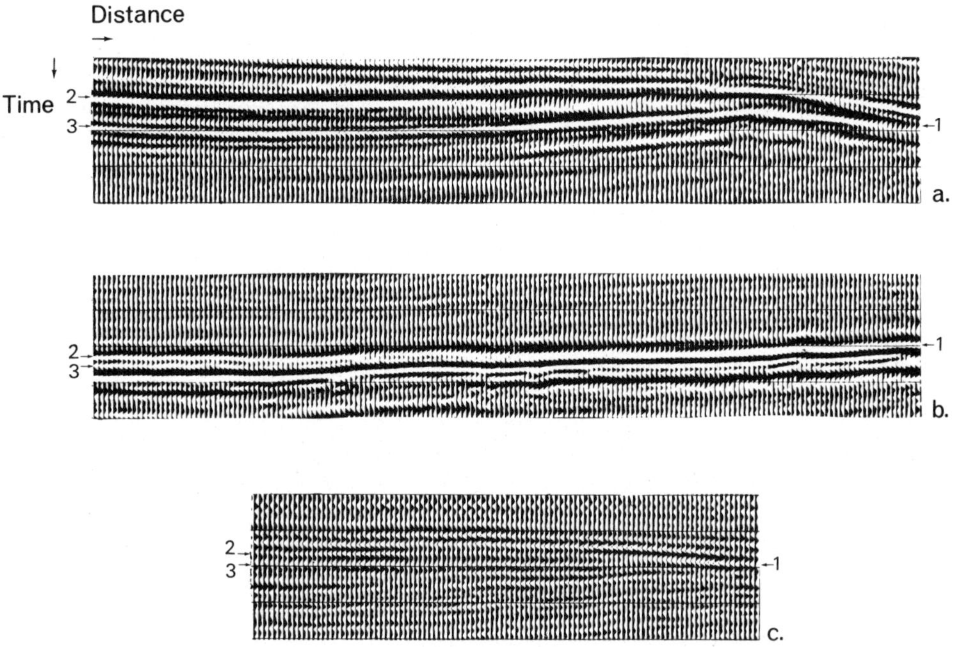

Fig. 5. 'Split' reflections.
An expression of change of thickness in part of the geological sequence.
Reflection (1) on the right-hand side becomes (2) and (3) on the left-hand side. In (5(a)) follow the black (peak) and in (b) and (c) follow the white (trough). Note that the lower horizon splits again in (5(a)).
In all three examples there is a discordance at the level of the horizon (1). In (a) and (b) it is a simple angular discordance. In (c) there is a gentle unconformity seen by looking along the reflections with the eye near the level of the paper.
The interpreter tracing horizon (1), and coming from the right is faced with a decision which is a purely geological one. If he is tracing horizons for the purpose of preparing structural maps, he must decide whether he wishes to portray the structure of the upper or lower formation: and he must be aware of what the decision implies, and report it correctly, or he must map them both. For example, if (1) is the boundary between Tertiary and the Chalk, selection of seismic horizon (2) may lead the interpreter on to the top of an incoming wedge of Palaeocene. If the circumstances are different, choice of seismic horizon (3) may lead him under an incoming wedge of Danian.
Which of these two situations exists can often be established by observing the overlap or overstep conditions. In (5(c)) horizon (2) appears to overstep the deeper beds: in (5(a)) the horizon (peak) above (3) appears to overlap the underlying material, and so does the horizon (peak) above horizon(2).
Thus, if the requirement is to map the formation boundary one would follow seismic horizon (3) in (5(a)), and horizon (2) in (5(c)). Such decisions would not be taken at single observations, as illustrated, but after examining the behaviour at all the places on the area to be examined where the split occurs. (5(b)) is an example of a very gradual split which does not provide any evidence on the nature of the boundary.
It will be apparent that whatever decision is taken by the interpreter must be taken consistently throughout the prospect.

then they have no counterpart in the subsurface. These non-geological features of seismic section are less common than geological ones, especially in present-day seismic sections. Their origins, and criteria for recognition, are discussed in later sections.

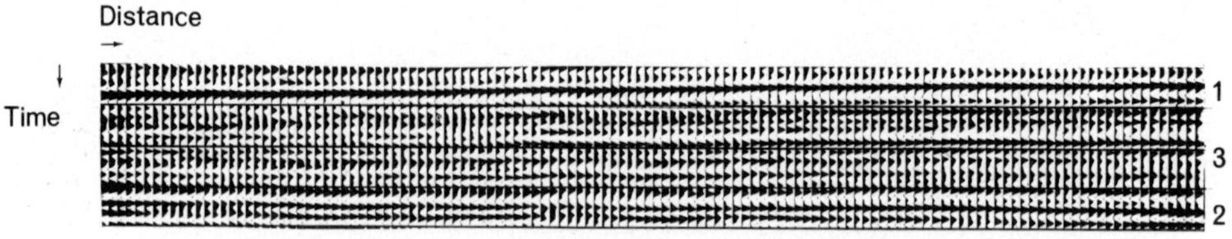

Fig. 6. An example showing very slow convergence of reflectors. Horizon 1 dips to the left, horizon 2 dips to the right. At about horizon 3 or a little lower the beds are completely flat.
This convergence, and the frequency changes which accompany it, are not severe enough to have any material effect on the appearance of the section. If it is viewed obliquely, it will be seen that every peak and trough on the right end can be identified on the left end of the section. If the process were to continue in either direction, however, the response of the complex system would eventually induce 'splits' in the reflections which have no stratigraphic significance, though they may indicate changes of thickness.
The interpreter should note that by inspection of this seismic section he cannot, with any confidence, interpret either the structure or the changes of thickness. The section could be interpreted in terms of structure and thickness change, on the assumption of constant velocity. The section could equally be interpreted in terms of changes of average and interval velocities, on the assumption of horizontal reflectors. The truth about the real earth may lie between these two, or may lie outside. This last comment implies that the dips could be the reverse of what appears on the section: and the sequence may be thickening in the direction in which it appears to thin.
Those who wish can examine the numerical aspects of this, using the following values.

Reflection time at top of figure	1.0 s
Reflection time at bottom of figure	1.4 s
Average velocity to 1.0 s	3000 m/s
Interval velocity 1.0 to 1.4 s	3300 m/s
Trace interval	100 m

This figure emphasizes the need for careful velocity studies, particularly in areas where the rocks are almost horizontal, and the possibility of facies change cannot be ruled out.

1.9 Planning interpretation

The previous section sets out some of the difficulties which follow when split reflections are encountered. Interpretation decisions are much less difficult where two reflections converge to yield a single reflection alignment.

For this reason it is a good plan to select mapping horizons in the deepest part of a basin, after a quick review of the whole area to establish that the chosen reflections have a wide development. From this starting point, trace the horizons as far as possible in the thinning direction.

Irregular developments in sedimentary basins will prevent this being done throughout a survey, but it remains a useful generalization. Where decisions are taken at a 'split', it is imperative that consistent decisions are made at every other point where that split appears in the cross-sections of the survey. Consideration will show that unless this is done the continuity will be carried on two different horizons, and a mis-tie will result when the continuity is followed round a closed traverse loop. Seismic correlation tests, applied to closely spaced cross-sections (i.e. at right angles to the strike) will often reveal any inconsistency of decision.

It is often the case that the control points where geological formations are identified are not in the deepest parts of basins. Such control points are often wells, most of which are drilled at structurally high locations where the sequence is more likely to be thinner than thicker. The control may arise from a neighbouring seismic survey, or occasionally by projections from outcrop observations; and these control points, also, may be far from the deep parts of the sedimentary basin. Even in these cases it is usually preferable to select horizons in the region of thick sedimentation and to trace them throughout the prospect. If there are other key horizons from the control points which are not represented in the first selection, it is always possible to add them within the established framework.

1.10 Naming of reflections

A consequence of the nature of seismic continuity (1.5) is that the rigorous identification of a seismic reflection requires *two* geological names: that of the rock above and that below the contact which generates the reflection. Each of them must have a litho-stratigraphic definition.

Determination of the geological position of a reflector in the geological sequence is achieved by extending the survey to include one or more holes drilled to the necessary depth, in which direct determinations of oneway travel have been made by lowering detectors into the hole and observing with a source at or near the surface. The naming can be carried, by the methods of continuity and correlation, into other parts of the survey.

Occasionally outcrop evidence can be extended usefully into the subsurface of the seismic section by the methods of constructing geological cross-sections.

In practice, naming is far from rigorous. Where continuity is carried on a reflection which marks a hiatus in sedimentation, new material may appear along the line of traverse to fill part of the hiatus from above or below. This new material may be a part, geologically, of the same formations as before – or it may represent a new subdivision to which a different name is given by the geologist.

Subsurface geological practice is to identify the tops of formations. It would seem reasonable to adopt the same naming system for the less rigid nomenclature of reflections.

For some reason seismologists sometimes identify reflections as the base of a formation. Sometimes this appears to be done because the character of the reflection is associated with the base of the upper formation, and is almost independent of what is below. For example, the base Tertiary reflection in some areas shows this characteristic. It results, probably, from two reasons. The Tertiary is often of notably lower velocity than any formation which underlies it. The reflection is perhaps generated by a characteristic group of basal Tertiary beds, and owes its character to them, and rather less to a Tertiary-Mesozoic contact.

Whatever the reasons, it would seem preferable to identify reflections with subsurface 'tops' wherever this can be done.

1.11 The roughness of reflections

In optics the roughness of reflectors is one of their identifying characters. Optically plane steel, lightly ground glass, specular hematite are all reflectors recognizable by the nature of the reflecting surface.

Roughness, in this context, means irregularities which are comparable with the wavelength of light. The corresponding definition of seismic roughness would mean irregularities of the order of a seismic wavelength. A 50 Hz seismic wave component travelling in a rock of compressional wave velocity of 3,000 m/s has a wavelength of 60 m. It is irregularities of this order which must be considered as seismic 'roughness'.

In nature, such irregularities are unusual, and the majority of reflectors are seismically smooth. Much larger irregularities are familiar: old topography, coral reefs, reflecting beds broken by step faulting are examples. These larger irregularities develop hyperbolas, as will be discussed later. They do not reduce the reflectivity by roughness.

Irregularities in the 60 m range are hard to find. Some facies of the 'argile scaliose' contain immense blocks which are in this range. Collapse structures resulting from salt removal or the collapse of limestone caverns may yield a jumble of rocks with this degree of roughness. These are unusual environment: and they are associated with geological and seismic complexities that over-ride the minor difficulties due to loss of reflectivity as a consequence of roughness of the reflecting surface.

1.12 A summary of the differences between geological and seismic sections

The seismic section is plotted in time, while the real earth is in depth.

The time-series is plotted vertically below the observation point whereas reflection point is at the end of a path which, after travel through the upper part of the earth, arrives at the reflector at normal incidence. The reflection point will, generally, not lie vertically below the observation point.

The seismic section contains elements which do not have a direct relationship with the reflecting horizons. They may be noise. Arrivals which are generated locally may appear quite random: noise from farther away may appear as alignments. There are, in addition, arrivals from the seismic source by modes of seismic propagation which are not compressional waves – shear waves and surface waves. There are compressional waves which are not simple reflections, but multiple reflections. There are arrivals from points which function as secondary sources in the subsurface.

Finally there are alignments or noise generated in the course of non-optimum recording or processing. Recording and processing cannot be optimum for every seismic arrival.

2 The geometry of seismic sections

2.1 Introduction

In 1.12 the fundamental geometrical differences between the seismic section and the real earth were pointed out. Detail shown vertically below the point of observation on the seismic section does not lie below the corresponding point of observation in the real earth. The points of detail lie at those points on the bedding planes at which the downgoing ray arrives normally, i.e. the wavefront arrives tangentially.

This leads to a difference between the seismic section and the real earth which can often be ignored in reconnaissance, but not in detailed work. The following sections describe effects seen on the principal structural features, anticlines, synclines and faults.

In this chapter, as in the last, the discussion is principally on 'ideal' seismic response and its interpretation. Little attention is paid as yet to noise and to other sources of corruption of the data.

2.2 Some features of anticlines on seismic sections

A gently folded anticline offers few problems to the interpreter. Fig. 7 shows the observation points on the line of traverse, and the corresponding reflecting points on the reflector. Since the seismic section presents its trace vertically below the observation point, it will be seen that the seismic section will portray the anticline as a little wider than it is in the ground, and the flanks a little less steep. The crest of the anticline, however, is in its correct position with respect to the surface, and the crestal region, with its low dips, shows the least distortion.

A strongly folded anticline shows these features more strongly, and the features become stronger with increasing depth of the structure, without any change of its real form in the earth. For example, a point on a flank dipping 30°, reflection time 1 s, velocity 3,000 m/s, is displaced 750 m. A similar flank at 6 s, velocity 3,000 m/s, is displaced 4,500 m. So if a sharp anticlinal roll with 30° flanks, 2 km wide, exists in the earth, its seismic representation, if it is at 1 s, will be 3½ km wide. If it is at 6 s, it will be 11 km wide on the seismic section. These are profound distortions. See fig. 7.

Because these travel times from the flanks of steep anticlines are so much greater than the travel times to the apex of the fold the reflection appears to dip steeply off to great depth, taking on a roughly hyperbolic shape.

It will be seen in later sections that these 'hyperbolas' can arise from almost any element of complex structure – anticline, syncline or fault. An important part of the interpreter's skill is his ability to unravel the complex patterns on sections which are made up of these hyperbolas.

It is of great value to consider folds in terms of their centres of curvature, as has been done, for example, by HILTERMAN (1970). Fig. 7 shows that the curve is a hyperbola, which has the shape which it would have if it were a diffraction from a point at the centre of curvature, i. e. the hyperbola is less curved than that expected at this time level.

Next, we consider the seismic representation of concentric folding: this postulates that beds have a common centre of curvature, and that they retain a constant thickness (measured normal to the bedding planes). It will be appreciated from fig. 8 that these reflections will appear as parallel folding – equal time intervals – on a seismic section. For a strongly folded anticline these will become a family of hyperbolas as illustrated in fig. 8.

On the other hand, if parallel folding exists in the ground, the seismic section shows the curious phenomenon of the fold increasing its width with depth as seen in fig. 8. It also increases its relief in the seismic presentation, so that a deep structure tends to look more attractive than the same real earth shape at shallower depth.

Pitching anticlines, where the pitching axis is at right angles to the plane of the section, can produce confused sections under some conditions. Consider, for example, the structure sketched in fig. 9 in which the reflecting horizons are horizontal immediately below the line of traverse, but to the north lies an anticline pitching to the south, the angle of pitch, and the slopes of the flanks, de-

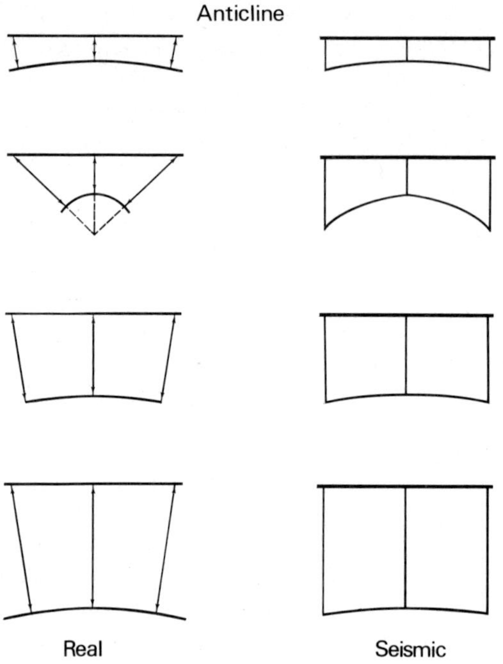

Fig. 7. In these sketches of real earth anticlines and their seismic response note
(I) That a gentle anticline is scarcely changed in its seismic expression.
(II) That a sharply-folded, narrow anticline may have as wide a seismic expression as a gently-folded, wide anticline. The latter will show less relief on the seismic response.
(III) A given anticlinal shape will have a wider seismic expression the deeper it gets.

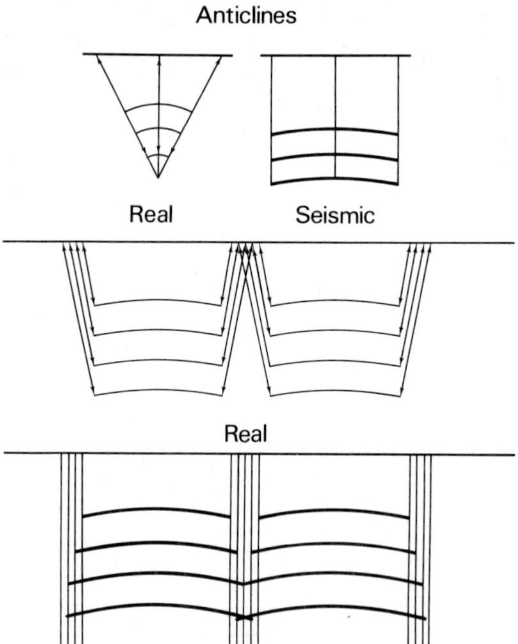

Fig. 8. Illustrating the seismic expression of concentric anticlinal folds as parallel folds on the seismic section.
Illustrating also the widening in depth of the seismic response of parallel folding in the real earth – a process which can progress to the point where neighbouring anticlinal folds can overlap in their seismic expression.

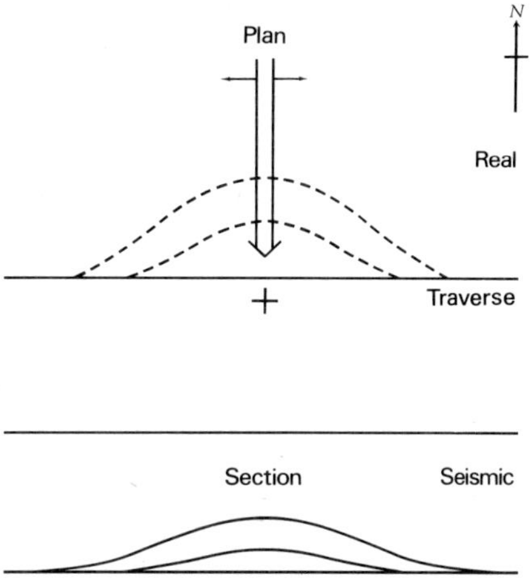

Fig. 9. A sketch illustrating the seismic response of a pitching anticline, with the angle of pitch decreasing towards the line of traverse, and the amplitude of the fold decreasing also, so that no folding is seen in a vertical real earth section along the line of traverse.
Provided that the increase in the pitch angle away from the line of traverse is sufficiently large, it will be found that there are several lines which can be drawn over the structure which are the loci of the feet of normals from the line of traverse. All of these lines (shown as dashed lines on the plan sketch) give rise to reflections on the seismic response, as shown.
It is unusual to find a single geological horizon fitting the geometrical requirements so exactly as to generate the pattern of the lower sketch. Rather, many reflecting horizons will fulfil the requirements locally. The seismic response is then a series of reflections of vaguely hyperbolic shape, overlying and interfering with each other.
These geometrical requirements are often met on traverses passing close to salt pillows or salt domes. These will often show concave upwards curvature on their lower slopes, and so generate the seismic response described.
It is important to recognize this response for what it is, and not to interpret the reflections as a salt feature, showing top and floor, lying vertically below the line of traverse.

creasing towards the line of traverse. The broken lines indicate the feet of normals drawn from the line of traverse, and show the origin of the three sets of reflections.

Diapyric folds, in which the limbs of the anticline both become vertical in depth, overturn, then return through verticality to normal dips, are rarely seen, and still more rarely encountered in seismic surveys. As always, the crestal part of the fold is presented with little distortion, while the steepening flanks appear as 'hyperbolas' extended to large reflection times. Below this the convergence of the beds in depth produces some of the features of a syncline, and the still deeper divergence of the beds produces some anticlinal features. These realtionships are illustrated in fig. 10.

Some reference must be made to anticlinal structures which are associated with intrusive bodies. The intrusive material may be salt, a mobile shale, or a volcanic rock. In many cases the beds are strongly folded around the tip of the intrusive body, and droopy hyperbolas obscure the flank structures deeper down. The intrusive bodies have a piercement relationship in most cases, so that the beds which are cut across show the typical fault features described later.

It is often difficult to identify the material of an intrusion. All one can do here is to review the classes of evidence which must be examined in a particular case.

Salt is usually flanked by 'peripheral sinks' or 'rim synclines', as described in chapter 6, while shale and igneous intrusions are not. In some occurrences the salt arises from a very deep source – such as the Cambrian salt in Iran – and the seismic resolution may not extend to such great depths.

Other geophysical methods, principally gravity and magnetic surveys, often provide the necessary diagnostics.

Unless the area of the survey is very remote, geological knowledge of the province is usually full enough for guidance as to the most probable nature of an intrusive body.

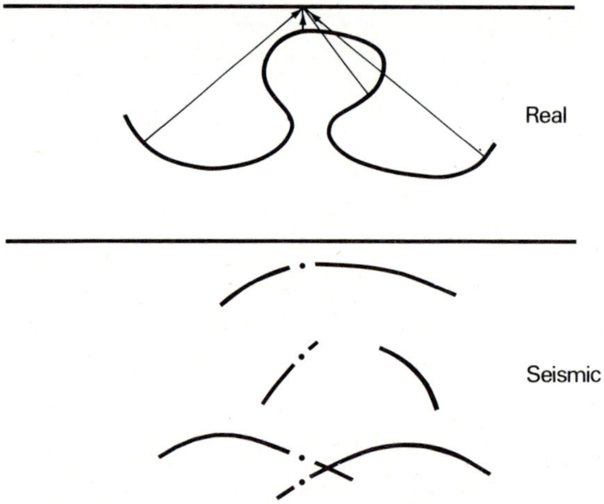

Fig. 10. Shows some of the features to be expected in the seismic response of a diapyric fold. Complex folds such as this yield a seismic response which is, at first, unrecognizable. Later, it will be shown, the seismic response of such a model as this can be drawn by first drawing normals to the reflector on the real earth model, as shown, and measuring the distance to surface. This distance, doubled, and divided by the velocity gives the normal reflection time. This reflection time, plotted to a convenient scale vertically below the surface point of intersection of the real earth model gives a point on the seismic response: and by an extension of the process the complete response can be lined in.
The response of the flanking synclines is treated in the next section. The failure to observe reflections from the steepest parts of the limbs is dealt with much later fig. 33. With a uniform velocity above the reflector in this figure such steep dips would be observed as far as noise level permits. The figure cited above, and those which follow it, will show why, in the real earth, they are sometimes screened, and cannot be recorded.

2.3 Some features of synclines on seismic sections

A gently folded syncline presents few problems. Fig. 11 illustrates that the syncline becomes a little narrower on seismic section than it is in fact: and that the lowest point of the syncline is undistrubed.

This simple relationship between the subsurface and the seismic section persists in all synclinal folds as long as the centre of curvature – or the locus of the centres of curvature – remains everywhere above the surface. If this condition is met, then steepening the flanks of the syncline, or lowering the syncline in the earth without change of shape, will cause it to contract in width on the seismic section.

Where the centre of curvature lies on the surface, it is seen that the syncline reduces to a single point, since there is only one observation point from which normals can be drawn. A series of cylindrical anticlines and synclines with the synclinal centres of curvature at the surface will show the anticlines as hyperbolas, meeting in cusps which are the locations of the one-point synclines.

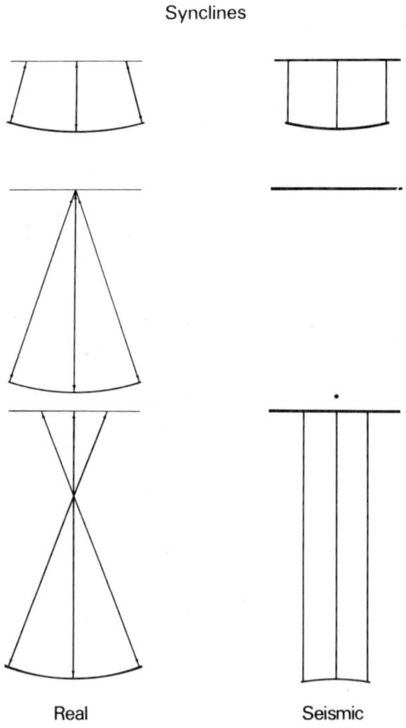

Fig. 11. Concentrically folded synclines appear as parallel folding in the seismic response (this relationship is not figured). With parallel folding in the real earth, the deeper the synclinal reflector considered the narrower its seismic response. A depth is reached, eventually, at which the seismic response of the whole syncline is a single bright point. Below this level, the normal to the *deepest* part of the syncline is *shorter* than the normal to any point on the flanks of the syncline. In other words, the seismic response of such a syncline is *convex upwards* – it looks like an anticline at first glance.
These relationships can be re-stated in terms of the position of the centre of curvature of a cylindrical fold in right section. Where the centre of curvature is above surface, the synclinal response is concave upwards. Where the centre of curvature is at the surface, the response is a point. Where the centre of curvature is in the subsurface, the synclinal response is convex upwards.
Centres of curvature in the subsurface are often described as 'buried foci'. The description is not very appropriate. The point is not generally the focus for an incident flat or spherical wavefront. It is simply the point of intersection of all the normals to the reflector: it exists strictly only for folds of circular cross-section, though most folds give a fair approximation.

Where the centre of curvature lies below the surface it will be seen that quite curious things happen: they are illustrated in fig. 11. First, the two anticlinal features, represented by hyperbolas, overlap. The cusp of fig. 8 has developed into a crossing point. Next, consider the syncline itself. The travel path to the lowest point of the syncline has the *shortest* length and the *shortest* travel time. In other words, the seismic presentation of the syncline is convex upwards, while the syncline in the earth is, by definition, concave upwards. It will be seen that within the syncline the reflecting points progress from right to left, as the observation points progress from left to right. Any minor features of the syncline will appear on the opposite side of the seismic syncline from what is found in the earth. A seismic section to illustrate this inversion is shown in fig. 12A, where the syncline in the centre is seen inverted in the lower part of the section.

Concentric folding in a syncline produces results rather similar to those described for anticlines. The appearance on the seismic section is that of parallel folding. The results are illustrated in fig. 13.

Parallel folding in a syncline produces strong changes with depth: these changes are an expression of the changing position of the centre of curvature as the deeper geological horizons are considered. Fig. 14 illustrates these changes, which are simply assembled from the previous figures.

Pitching synclines, where the axis crosses the line of traverse, can produce the most complex regions on seismic sections. There may be several normals from the line of traverse to the pitching axis, and the radii of curvature at right angles to the axis may be different on the several normals –

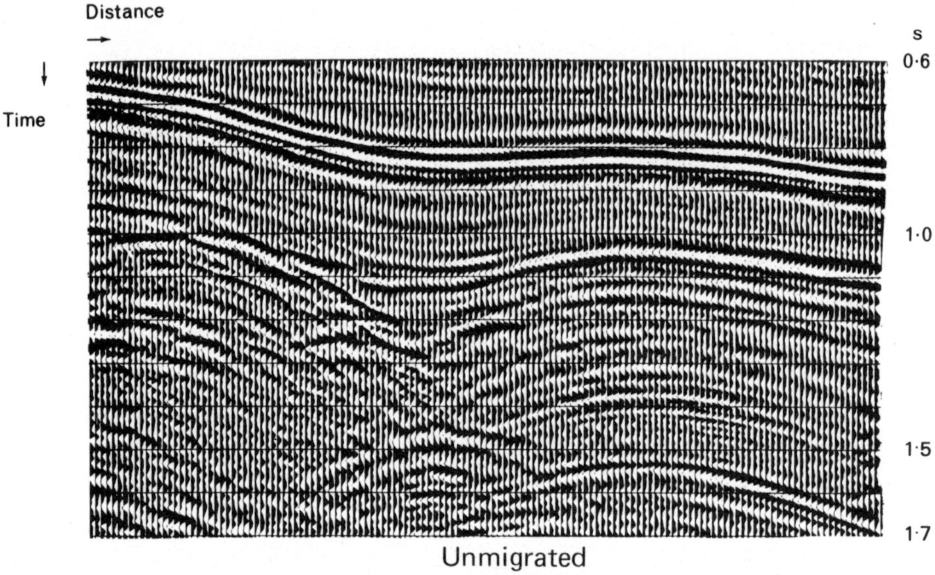

Fig. 12 A. The middle part of this cross-section is synclinal – a low amplitude fold at shallow levels, increasing in amplitude downwards. It shows well a wide syncline at 0.8 s (measured between points of inflexion). It is narrower at 1.1 s, still narrower at 1.2 s. At about 1.4 s it is represented by a point, and below that by a convex upward reflexion. Thus all the styles of syncline expression are included.
Several other useful observations can be made from this section.
In a region where all the fold responses are convex upwards, one needs some criterion to distinguish anticlines from synclines. It is that the 'hyperbola' at a given time level is strongly curved for a syncline, gently curved for an anticline. In other terms, the 'focus' of an anticline reflector must be buried deeper than that of a syncline at the same reflection time. This is well illustrated in the time interval 1.4 to 1.6 s.
Consideration of a real earth model will show that a normal geological relationship along the locus of horizontality of the bedding is preserved in the seismic response. In other words, the seismic response along this locus on the seismic section is just the same as if the beds were horizontal – as, locally, they are. This is true whether the locus is the axis of an anticline or a syncline.
Hence, however confused the seismic section may appear, it is still the case that valid seismic correlations can be made between the core of an anticline and the core of a neighbouring syncline. Reference to the figure will show that it is difficult to correlate between the syncline and the right-hand anticline in this case: but this question will be reconsidered under fig. 12 B.
Finally, between 1.2 and 1.3 s, close to the synclinal axis, are two convex upward arrivals with strong curvature, looking like a pair of eye-brows. These can only be interpreted as two small synclines – an interpretation difficult to appreciate in this structural context. This point will also be referred to under fig. 12 B.

a situation illustrated in fig. 15. The resulting seismic section can show alignments overlying each other which are gently concave upwards, convex upwards, and steeply intersecting.

It is the seismic expression of synclinal features that contributes most to the difficulties of complex areas. A few complex geological environments are discussed in a later section.

2.4 Some features of faults on seismic sections

Geologically a fault is a discontinuity, seismically it is not. The reflected wavefront does not have terminations: it is distorted by the fault. In complex situations it may develop secondary fronts; shad-

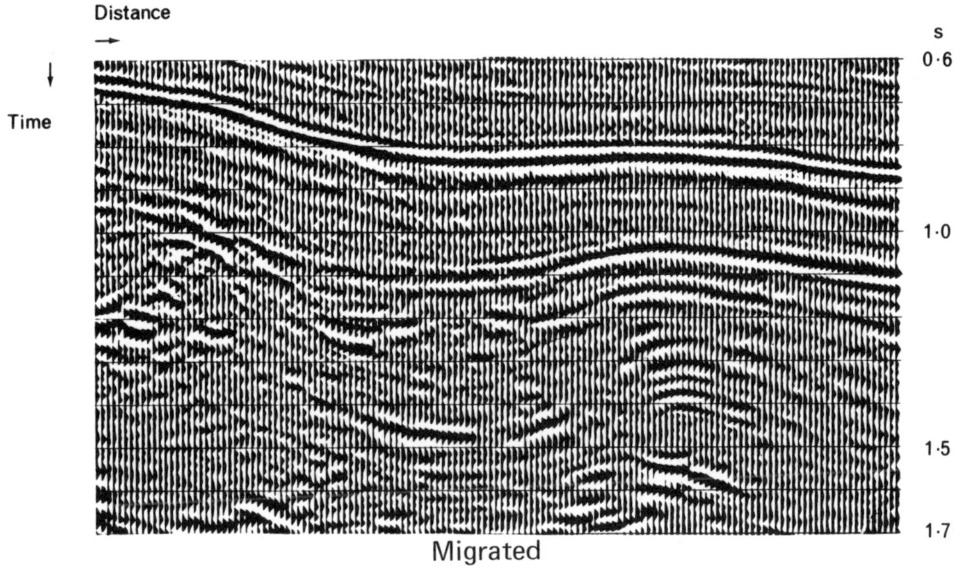

Fig. 12 B. This is a migrated version of fig. 12 A (see Level 2). For the present one can regard this as close to the real earth model whose seismic response is fig. 12 A.
Points to note are:
The syncline below 1.4 s has the shape that one would expect to find.
The reflection at 1.3 s at the crest of the right-hand anticline is that at 1.47 s in the trough of the syncline. With this guidance it is now possible to return to fig. 12 A and establish the anticline – syncline correlation.
The 'eye-brows' of fig. 12 A are now seen to be two small synclines, with a hint of a small anticlinal roll between. This could be a small salt body, or a mobile shale.
The right-hand anticline is seen to be complex, and possibly faulted in depth.
The steep limb between anticline and syncline has very little expression. This could result from the phenomenon described in Level 2 fig. 33. It could equally well result from a failure to scan enough traces in the migration process.
The reader can return to this figure after reading Level 2. Observant readers will already have noticed a change of phase between figs. 12 A and 12 B. To confirm, look at the horizontal reflection at 0.8 s – it is white on fig. 12 A, black on fig. 12 B. Similar observations can be made everywhere in the core of the anticline or the syncline.
This is, of course, a mistake. It is retained to show how the work of the interpreter can be complicated by quite small errors of processing or presentation. The reader may have had difficulty in using this figure to establish the anticline-syncline correlation on fig. 12 A. It results from this phase change.

ow zones may develop. The seismic expression of the fault on the cross-section is a record of the arrivals from this distorted wavefront, complete with secondary fronts and shadows. It is in the nature of wave fronts to 'heal' as they progress through the earth: the distortions become less marked– small faults at great depth tend to be slurred over on seismic sections.

Geological features of faults include the essential discontinuity: a break across which there is a displacement of the rocks on one side with respect to the other. Some faults, in hard rocks at shallow depth, will show little else, but in most cases other features which affect the seismic expression are to be observed.

Fig. 13. A sketch of parallel folding: two anticlines and a syncline.
This is a diagrammatic presentation of the structure elements of fig. 12 B.

Folding is frequently associated with faulting, particularly in softer rocks and at depth. An often observed circumstance is the occurrence of a well-defined fault system in basement rocks; renewal movements, if they are gentle, will produce little flexures in overlying soft rocks. As these movements become stronger, they will produce flexures faulted in the middle limb, either throughout their course or locally, and either fault or flexure may dominate the structure.

Mechanical changes, and sometimes lithological changes, take place in the rocks on either side of the fault. Rocks are crushed or distorted. Joint planes are opened up. Circulating meteoric or juvenile waters, or other fluids, have an easier passage through the rocks near the fault than elsewhere, and a wide range of chemical changes can result.

The fault plane itself is not a plane in the geometrical sense. It is a geological entity whose thickness may be significant. Its lithological constitution can be very varied. Under some conditions it may be a gouge of finely comminuted rock material with perhaps large blocks of the harder rocks embedded in it. It may be filled with sediments from higher horizons if the faulting takes place under tensional conditions – as in the case of sandstone dykes. Finally, quite foreign materials can be introduced into the fault plane. These may be veins of quartz or calcite, lodes of metallic minerals, or igneous dykes of any width.

The seismic features of faults are the essential displacement of reflections on one side with respect to those on the other, and some associated features. These associated features are three – though the development of any one of them is entirely dependent on the local circumstances.

These features are 'hyperbolas', reflected refractions and fault-plane reflections.

The hyperbolas may have the curvature appropriate for the T_0 value (at the vertex) and the corresponding $\bar{V}$ value: or the curvature may be greater or less than this. These features can be interpreted as anticlinal or synclinal folds of diffracting points, taking into account the angle between traverse and strike, using preceding descriptions. There may be a phase change in the hyperbola, or there may not. The amplitude of the hyperbola may be comparable with that of the associated reflection, or it may be less.

Fig. 14. The seismic response of the model of fig. 13. This brings together the features of figs. 7 and 11. The anticlines widen in depth, until they overlap. The syncline narrows to a cusp, and deeper it assumes a convex upward expression.
Anticipating the results of some later sections, one must point out that the truncated ends of reflectors at a fault plane also act as 'buried foci', and give rise to hyperbolic arrivals on seismic section.
The figure illustrates one type of interpretation error which can arise if these geometrical relationships are not appreciated. At the shallowest levels the structural elements will be correctly portrayed, although the anticlines are too big and the synclines too small. At intermediate levels the interpretation might be two anticlines separated by a fault. This is not very plausible geologically, and is a suspect feature on any seismic interpretation. Finally, at greater depth, the seismic section may be dominated by a great convex upward arrival from the syncline which obscures the anticlinal arrivals. This synclinal arrival may be of higher amplitude than the anticlinal one because of the focussing effect of a concave reflector and the de-focussing effect of a convex one. The mistake which can be made here, if the higher developments are not seen, is to map an anticline.
Anticipating some later results, the identification of these hyperbola-like seismic arrivals depends ultimately on identifying the 'buried foci' or centres of curvature that give rise to them. This is done by considering the degree of curvature of the arrival: and as will be seen later this is a function of the velocity.
For the present it is enough to say that for

Anticline	Centre of curvature is below reflector
Fault	Centre of curvature is at reflector
Syncline	Centre of curvature is above reflector.

Reflected refractions may indicate refractor velocity and structure in a bed which has been faulted: or they may indicate these features of some higher high-velocity bed that functions as a waveguide.

Fault-plane reflections are recorded most frequently from fault planes with low hade.

In some environments the only seismic phenomenon to be observed in association with a fault is a zone of noise, in which no clearly identifiable seismic arrivals can be recognized with certainty. At this point one should warn against the recognition of very short alignments. With only two elements, black and white, the laws of chance will produce short alignments from time to time in a welter of random noise.

Geological and seismic features of faults have been separately described. It would be logical now to establish the exact correspondence between them – to establish an interpretation code which would take the interpreter from his observed seismic phenomena to the corresponding geological configuration. This cannot be done with any completeness, for several reasons. Geological observation of fault detail is surprisingly incomplete. There are good outcrop descriptions and photographs, though these are sometimes subject to surface disturbances, and velocity measurements are not made. Subsurface petroleum exploration does not give the fine detail in the neighbourhood of faults – at intervals of 20 m or so – which is required. Faulting is sometimes a matter of inference. Even mining operations yield less information than one would like. Coal miners dislike working in faults for safety and commercial reasons.

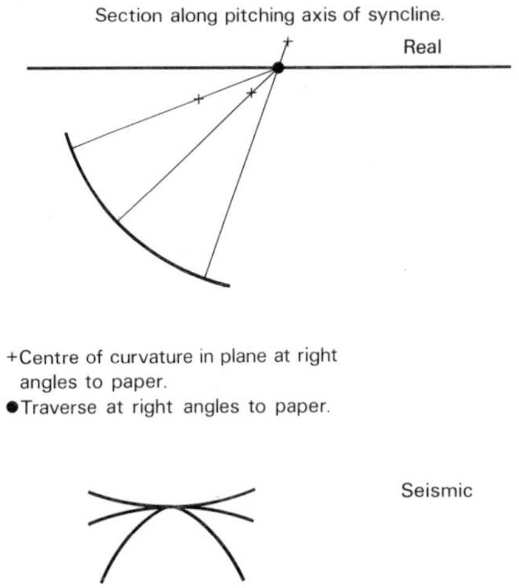

Fig. 15. As in the discussion of fig. 9, it is seen that the pitching syncline can generate a complex system of arrivals if the angle of pitch is increasing so rapidly away from the line of traverse that several normals can be drawn from a point on the traverse to the axial region of the fold.
An added complication, shown in the upper sketch, is that the centre of curvature of the fold, as seen in cross-section, may change its relationship to the surface. This upper sketch is a section taken along the pitching axis of the syncline, at right angles to the line of traverse. The crosses indicate the centres of curvature of the fold in sections at right angles to the plane of the paper.
The lower sketch shows the confusion that arises when these several arrivals are shown on seismic section.
Geologically, the geometrical shape is not very common. A few seismic examples have been found in the secondary peripheral sinks around salt domes. In places these are completely annular in form, even if irregular. Elsewhere the development is asymmetric, and the sink is lune-shaped in plan. The ends of these lune-shaped sinks present the configuration required, and some examples of the seismic response of the lower sketch have been seen.
Other examples have been seen from very deep reflections below the ocean deeps. Extremely gentle folds at 10,000 or 12,000 m are enough to provide the configuration of the upper sketch.

One result of the lack of readily available geological descriptions has meant that realistic models have not been available for seismic modelling studies. Considerable progress has been made in the study of the simplest model of a clean fault displacement in papers by HILTERMAN and TROREY: though even in this case a full wave-equation treatment is not yet available.

Fault interpretation is thus much less securely based than fold interpretation, since folds are known with precision from drilling results and surface geological studies. What follows is the best guidance that can be given: but the reader should be prepared to question and re-examine in the light of this lengthy preamble.

Fig. 16 illustrates the simple configuration which yields little seismic expression other than the displacement of the reflectors.

More usually the simple configuration is that of fig. 17, and this, according to TROREY, leads to continuation of the reflection as a diffraction hyperbola. This hyperbola is tangential to the reflector at the point of truncation. The backward continuation of the hyperbola has the reverse phase, the phase change taking place at the point of truncation. The phase of the hyperbola is determined by the velocity relationships. If the observer sees the same relationships between high and low velocities when looking obliquely at the truncated end as at the reflecting surface, there is no change of phase. If the velocity relationships are reversed, then the hyperbola is of the opposite phase from the reflection.

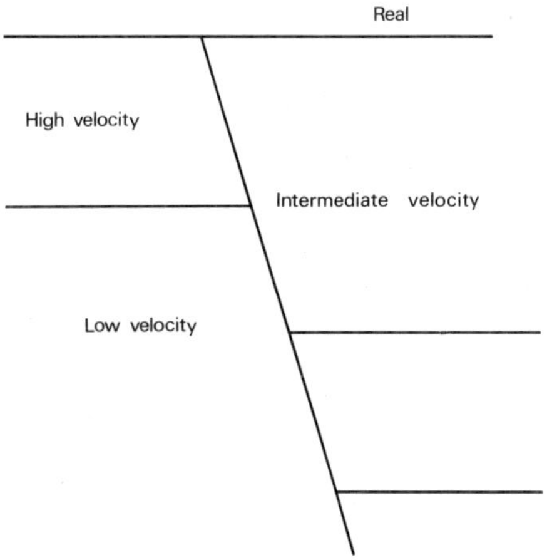

Fig. 16. Sketch of normal fault.

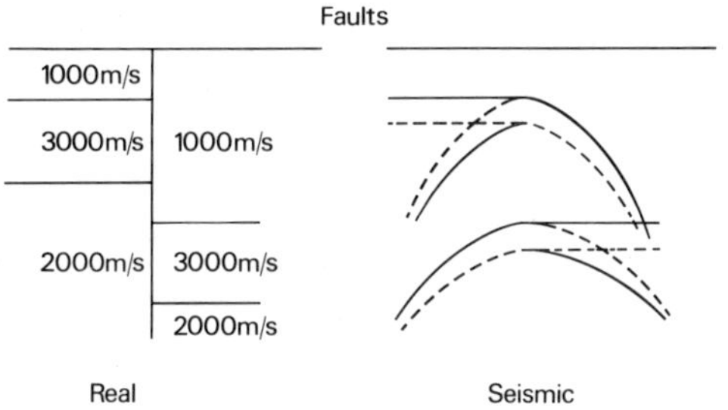

Fig. 17. Sketch of normal fault and its seismic response. Reflections, and hyperbolas of a positive polarity are shown solid, those of a negative polarity are dashed.
Notations on the real earth model refer to the seismic velocity.

Fig. 18 illustrates these relationships, which can usually be found in any large survey. Often the necessary detail is obscured by the reflections themselves. Often the change of phase is demonstrably not present. Amplitude relationships do not conform with the available analysis.

It can easily be established that the diffraction hyperbola is tangential to the reflection at the point where the reflection ends. The vertex of the hyperbola, which is always up-dip from this tangential contact, represents the true position, i.e. the migrated position, of this truncated end of the reflector in the real earth. It is this relationship which justifies the opening sentences of this section. The course of the reflection carries it smoothly into a diffraction. Seismically there is no discontinuity at a fault. Wavefront constructions will demonstrate that this is the case.

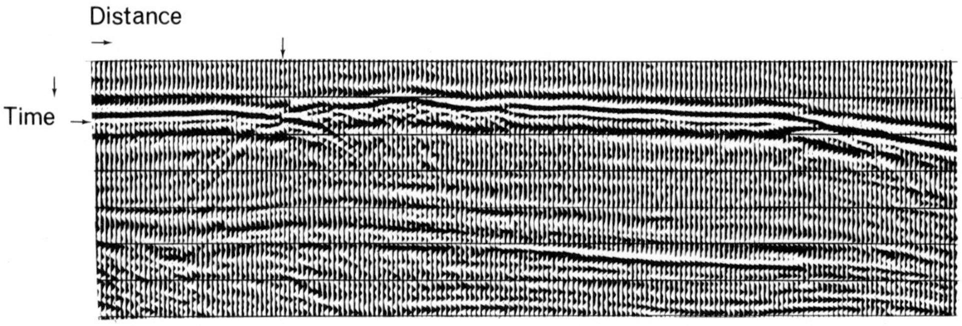

Fig. 18. Section showing a series of small faults on an upper horizon. They cannot be identified at lower levels, and may not exist there.
The left-hand end shows what is probably a classical change of phase between reflection and forward hyperbola, and between forward and backward hyperbolas. Time and trace are indicated by arrows.
Note also the alternative interpretation. A fault has two sides, and what appears to be a forward hyperbola from one side of it may equally well be a backward hyperbola from the other. The change of phase may be merely a coincidence.
What is probably observed is the addition of forward and backward hyperbolas on both sides of the fault. Neither observation nor analysis has yet much to contribute to the precise identification of such fault events. Perhaps identification with a geological element is not a meaningful phrase.
Identification of such small faults, and of the unbroken ground between them, is of great importance in coal mining geophysics.

An important practical observation at this point is that it is extremely confusing to mark on the same section or map the seismic section positions and real earth positions of any feature. If migrated positions are of interest (they usually are), they should be shown on separate maps or sections.

Fig. 19 illustrates also what may happen where the reflector is a strong one, with a positive reflection coefficient, i.e. with a low velocity rock overlying a high velocity rock. As the observation point moves further from the fault the angle of the direct path from the diffracting point with the reflector eventually falls below the critical angle. At this distance there are two arrivals. The direct one is a wide angle diffraction. The second is a reflected refraction – refracted in the high velocity rock, reflected from the fault face.

These two features are readily distinguished on the seismic section. The direct arrivals are represented as a diffraction hyperbola. The reflected refraction is a straight line arrival, whose slope is dependent on the velocity in the lower bed and the dip. Any structure in the refractor bed will appear as departures from the straight line.

The reflected refraction will be tangential to the diffraction at the critical distance: and larger amplitudes can be expected in the neighbourhood of the critical distance. It will be clear that this reflected refraction can appear only on the backward-directed hyperbola. In the forward direction the refractor does not exist.

Still another fault arrival is a simple reflection from a fault-plane. It is necessary that the elastic properties of the two rocks separated by the fault differ sufficiently strongly. For the fault reflection to be traced very far the contrast must be maintained along the fault. If this condition is not met,

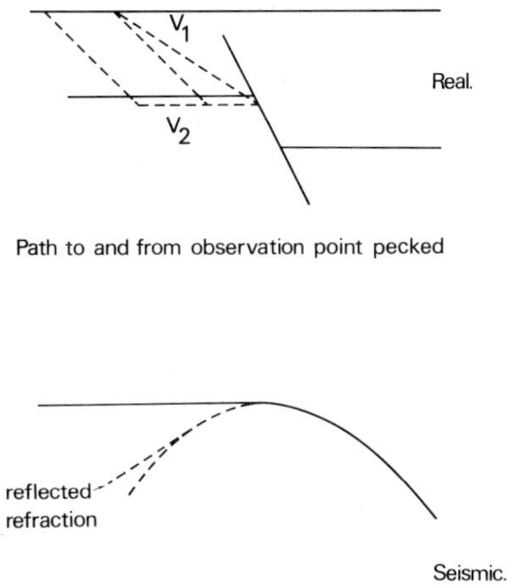

Fig. 19. Upper sketch shows a normal fault, and a possible reflected refraction path if $V_2 > V_1$.
The lower sketch shows the seismic section with this reflected refraction tangential to the diffraction hyperbola from the fault.
Two points are worth adding:
(I) The reflected refraction can exist only with the backward hyperbola. On the side of the forward hyperbola the refractor does not exist at this level.
(II) The reflected refraction contains the same structural information as the reflection itself. Distinctive features of one can be found on the other. This will often help to confirm the identification.
Some dependence of amplitude on the hade of the fault could be expected: it has not been observed.

the fault-plane reflection will vary so much in amplitude and character that the continuity will be impaired. Some fault-plane reflections are so persistent in continuity and so regular in character while cutting a great thickness of varied sediments, that one is led to the conclusion that the fault-plane itself has lithological characteristics which contrast strongly with those of the enclosing rocks. This would be expected if the fault action produced a significant thickness of broken rock material or gouge, especially if this material became water-saturated or gas-cut: such changes would produce low velocity fault material. High velocity definition of a fault can arise from the formation of 'sandstone dykes', but much more frequently from vein formation, with calcite or dolomite in the fault fissure. Mineralization with base metals could similarly give high velocity character to the fault-fill, and so could igneous rocks intruded into the fault. Such phenomena have not been well observed and recorded by seismologists, nor checked against the subsurface geology.

Fault-plane arrivals may be obscured by reflections from bedding planes and by diffractions. They are sometimes seen clearly when reflections and diffractions are rather poorly developed.

The determination of the hade of a fault, the discrimination between normal and reversed faults, is not always easy on seismic sections. The apparent terminations of reflections are not sufficient evidence. The neighbourhood of the fault trace on the seismic section is usually obscured by 'noise' made up of diffraction hyperbolas. Strong reflections, usually the upper ones, can be traced further into this zone than the weaker ones, usually the deeper reflections.

The acceptable evidence is derived from the hyperbolas. The vertices are in the real earth position with respect to the surface, and represent real earth points on the fault. If the locus of vertices in a system of fault hyperbolas establishes a slope, this is the hade of the fault.

A still more difficult distinction is that between a sharp flexure and a fault. Such a flexure (concave upwards, or decreasing the dip in the downdip direction) on a monocline produces similar geometrical relationships to a sharp syncline – which is readily demonstrated by tilting fig. 18. The controlling factor is again the position of the centre of curvature. If the flexure is strong enough to bring the centre of curvature below the surface, then the 'horns' of the hyperbolas appear, and the flexure looks like a fault.

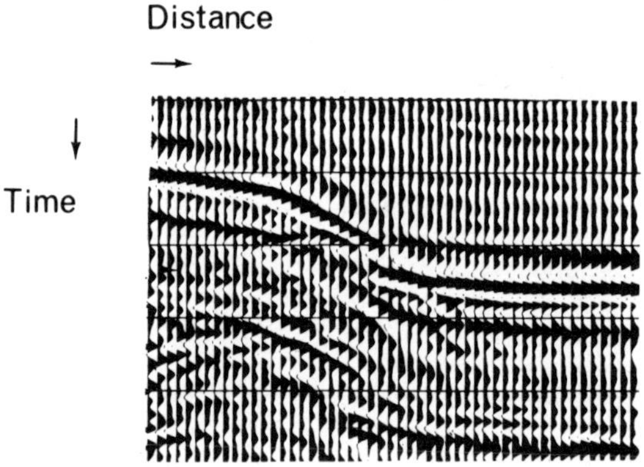

Fig. 20. A small flexure. Points which distinguish this from a fault.
(I) The hyperbolic forms appear to terminate somewhat abruptly (as in fig. 12 A) rather than to become progressively feebler, and be lost in the noise and other reflections. This is a somewhat subjective judgement.
(II) The hyperbolas seem to have their vertices arranged along two parallel axes – not a single axis, as they would be along a fault plane.
(III) No phase reversals in the forward hyperbolas can be observed.

It will be shown later that discrimination between fault and flexure requires accurate velocity and strike data, and possibly migration, for the most difficult cases.

Sharp flexures often appear in pairs close together; one convex upwards, the other concave upwards, with the two separated by a steep middle limb. Such a structure often appears in beds above a fault in older rocks which has renewed its movements at a later date, but where the movements have not been sufficient to disrupt the upper beds. In such a double flexure the centres of curvature for the convex upward and the concave upward flexures are often close together for any bedding plane involved. If the steep limb of the flexure is vertical, then the hyperbolas will continue theoretically to infinity on the steep limb side: in practice they will continue until they are lost among seismic signals of higher amplitude, or in noise – or terminated by the limits of the recording or of the seismic section. Where the steep limb does not become vertical, it is seen that the horns of the hyperbolas will be terminated. The 'horn' cannot be observed further along the traverse than the furthest surface point reached by a normal to the reflector in the steep limb of the flexure. In these circumstances the hyperbolas will terminate abruptly.

Figs. 20 and 21 illustrate a small flexure. There is an apparent break in the reflections of fig. 20, which at first suggests a fault interpretation. However, the vertices of the hyperbolas in fig. 20 fall on two alignments, not a single alignment as would be expected with a fault. This suggests the configuration of fig. 21. Moreover, the hyperbolas of fig. 20 appear to terminate abruptly:

this is a feature of fold-generated hyperbolas: the limit is set by the maximum dips involved in the flexure.

A special problem can be considered: the distinction between such a flexure with a steep middle limb and a fault.

Both of these configurations generate hyperbolas on a seismic section. In both cases the hyperbolas extend on either side of the feature.

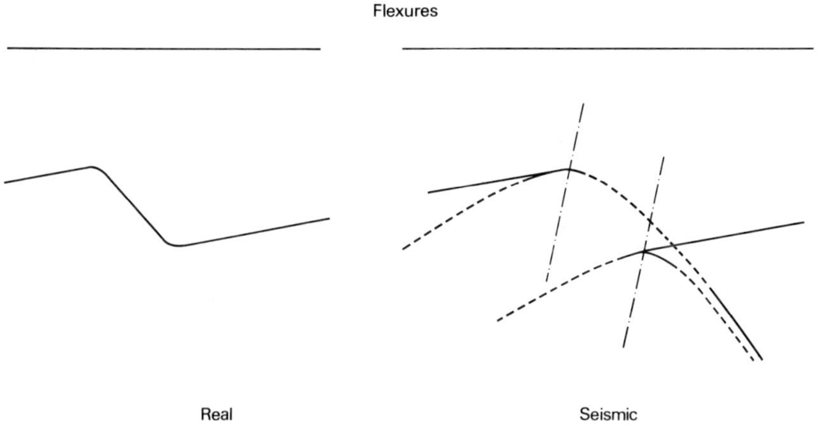

Fig. 21. Sketch of the conditions at a sharp flexure.
Note that the centres of curvature are very close to the reflector, so that criteria based on their relative position cannot be used to discriminate between anticlinal and synclinal elements of the flexure. Instead, one relies on correlation from one side to the other, and reconstruction of the middle limb, which is rarely recorded.
The geometry of this fold is such that the phase of the hyperbolic developments is the same as that of the reflec tions with which they are associated.

There is no established theory, checked against field observation which offers a full solution. If velocities and strike are known with sufficient precision, and the section is of high quality, calculation of point by point migration may show whether faulting or flexuring is present.

There is a further criterion. In the case of a fault which has only a narrow fault zone, the points at which bedding plane reflections are tangential to their associated hyperbolas will fall on the same alignment, irrespective of the side of the fault from which the reflections originate. Similarly the vertices of the hyperbolas from the two sides of the fault are colinear.

In the case of a sharp double flexure, the two axial planes develop their own systems of hyperbolas, and the points of tangency fall on two nearly parallel alignments. Similarly the vertices of the hyperbolas fall on two alignments.

This feature also can be seen on fig. 20.

Overthrusts are often difficult to recognize on seismic sections. The complexities of structure associated with them may generate wide areas of interfering arrivals. The differences of structure at higher and lower levels in the section might occur across an unconformity. The interpreter may not be able to make a firm decision from the seismic sections: but a knowledge of the geological environment in which he is working will guide him.

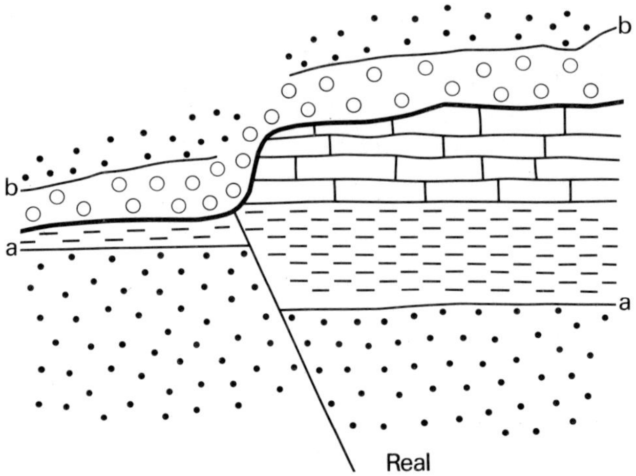

Fig. 22. An old fault-line scarp covered by later sediments. Note the possible confusion on seismic section when the reflections from a and b are correlated across the fault.
The geometry of this situation can be matched by two quite unrelated fault geometries. One is that figured, which shows a fault line scarp, and the draping effect of the immediately overlying sediments, continuing the fault plane upwards a short distance. The puzzling feature for the interpreter is that some horizons such as a, will appear to be down-faulted by a normal fault, while others, such as b, will appear to be reverse faulted.
The alternative is a blatte fault, one in which the movement is largely horizontal, and which brings a thicker development of the geological sequence against a thinner development of it – a basinal sequence against a more marginal one. This geometry, also, will show some horizons going down, others going up, on the same side of the fault.
Discrimination between these two cases cannot be attempted with confidence on a single section: but it can usually be achieved when a whole survey is available.
The fault-line scarp will appear always on the same formation, or perhaps a small number of resistant scarp-forming formations. These can be recognized by reflection character above and below those formations, and by characteristic velocities – usually high velocities. In some cases the same formations will be known at outcrop, and scarp formation under contemporary erosion conditions can be observed. In a fault-line scarp the change in apparent throw is always found between the same two reflectors (a and b), lying above and below the scarp-forming horizon – or a corresponding pair for each scarp-forming horizon.
Blatte- or tear faults show a similar change of thickness from one side to the other. In this configuration the neutral line, or horizon of no apparent throw, will vary from place to place, dependant on the vertical movement which accompanies the dominantly lateral movement, or dip variations from place to place. By contrast, the neutral line for the fault-line scarp passes always through the scarp-forming formation.

Faults with strong lateral movements will have no seismic expression unless the fault movement is sufficient to bring together sequences with marked differences in facies or thickness. Correlation across the fault then becomes uncertain. However, differences in facies and thickness can arise equally well across faults which are active with purely vertical movements during the course of sedimentation. The fault scarp can separate two palaeo-environments, and hence two facies of sedimentation. Where the fault scarp is less prominent, the environments on the two sides may be almost the same, but the rates of sedimentation may differ.

Again, it is a knowledge of the geological environment that guides interpretation of such a fault.

There are some examples of bold fault scarps that leave their record convincingly on the seismic section. Fig. 22 shows the course of development. The first attempts at correlation across the fault give contradictory indications of the direction of throw.

In all of the fault illustrations so far it has been assumed that the line of traverse is set at right angles to the strike of the fault. Fig. 23 illustrates the relationships when the traverse is oblique to the fault.

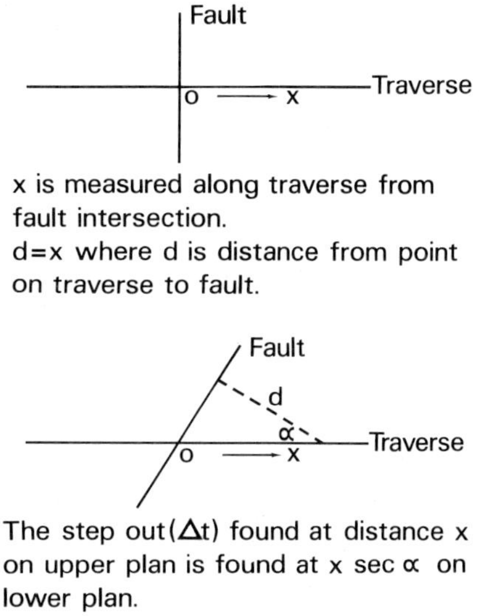

Fig. 23. The effect on the seismic section of running a traverse oblique to the full dip. Any hyperbola, developed from any 'buried focus' becomes wider with increasing obliquity of the traverse line.
Perhaps the easiest visualization is that the three-dimensional seismic response of a truncated reflector of a fold feature is a hyperbolic prism. Cut at right angles this gives the narrowest or most strongly curved section. Increasing obliquity gives increasing width.
It follows, also, that any traverse parallel with the strike of faults, flexures or folds will record apparently horizontal arrivals from these features. It is not unusual to find in a survey that the dip lines are clear, and can be readily correlated from one to the next, while the strike lines connecting them may be obscure, and may even suggest continuity at variance with the dip-line correlations. In a number of cases the effect can be shown to result from these near-horizontal arrivals.
The best solution appears to be to record and interpret dip lines first, and then to site the strike tie-lines where they will not intersect dip-line hyperbolas, if that is possible.

If the traverse had been at right angles to the fault, then the travel time would have been a function of x, the distance from fault intersection to observation point. Where the fault intersects the traverse at an angle α, the travel time is the same function of x cos α. In other words, to find the same travel time on the hyperbola from an oblique fault as at x on an identical fault at normal intersection, then one must observe at a distance x sec α; the greater the obliquity of fault to traverse, the more its hyperbolas are stretched horizontally, or flattened in appearance.

The reversal of phase between the forward and backward branches of the hyperbola at the point of tangency to a truncated reflector has been described and illustrated. It appears fairly often as far as one can see on cross-sections.

There is, however, a body of evidence for hyperbolas whose branches do not change phase in this setting. Some experimental work shows simple hyperbolas at the termination of a plate. Some cross-sections show hyperbolas which are associated with faults but show no change of phase.

What is not clear from the literature is whether this difference in seismic expression can arise simply from change of parameters in a simple model. The alternative to be considered is that the model is different in the two cases.

In those ultra-sonic experiments in which wavelengths are not properly scaled the effect of a slight burr on the end of the plate would be to create centres of curvature which would certainly generate hyperbolas with no phase change. In real earth faults it is frequently seen that beds are pulled down or pulled up into a fault-plane, thus creating sharp centres of curvature, or steep, truncated ends of beds. These configurations, also, would give rise to hyperbolas with no phase change.

Thus the true seismic expression of fault phenomena, and the detailed interpretation of what is observed, remains unsatisfactory until further work is done.

There is a further change of phase observation which needs more theoretical and practical work to confirm it. Occasionally there appears to be a change of phase in the hyperbolas associated with a sharp flexure. It would seem that this can arise in cases where the dip of one limb of the flexure is steeper than some of the ray paths to surface. Effectively, these rays see the underside of the reflector, with a reversal of the sign of the reflection coefficient. This relationship is not established. The nature of the flexure is often difficult to observe.

2.5 Interpretation of seismically complex areas

From the earlier sections of this chapter it will be clear that where the interpretation departs from the 'obvious' character of 1.2 it is because anticlines, synclines, faults and flexures have generated hyperbolic forms on the seismic section. Interpretation in these complex areas becomes a matter of recognizing the hyperbolas on seismic section, and replacing them by the points or circles of the real earth which generated them. The techniques of tracing continuity (1.7) no longer yield results which are geologically recognizable. Correlation, however (1.6), can still be applied, though often with difficulty, to the vertex regions of hyperbolas.

First one must examine more closely the 'circles' of the real earth, and the 'hyperbolas' of the seismic section to which they are transformed.

Not much has been written on the shapes of folds. There is very early work, before the use of seismic survey, in which the assumption of circular folding was made. Surface dips were measured by the geologists, and the assumption of circular folding was made to construct the shape of folds in depth. The results were close enough to the real earth shapes to discover oil in regions of strong folding that still pose problems for the seismologist.

More recently it has been shown that folds, up to the severity of folding found in metamorphic rocks, can be represented very closely by a simple sine wave, referred to oblique axes in extreme cases, with few harmonics. For a quarter-wave of a fold the match is very good.

Within the approximations usually adopted one can say that a real earth circle transforms into a seismic hyperbola, and vice versa. Circular arcs become hyperbolic arcs.

Within the same approximations it can be seen that the time differences in any hyperbola on a seismic section are those due to a hyperbola expressing the NMO at the vertex time (if the event is a diffraction), or the NMO at the time of the centre of curvature (if it is a fold). These observations are valid if the strike is at right angles to the traverse: if not, then x must be multiplied by the sec α term of 2.4.

A possible approach, therefore, to unravelling the problems of such complex areas is to use a plot of NMO at short interval times – say every 0.1 s – on a template to seismic section scale. By applying this to every recognizable hyperbola on the section one can mark all the vertices of hyperbolas (at the minimum reflection time), and the centres of curvature corresponding to them (at the time of the fitting hyperbola, and on the same trace).

Such a plot shows all the elements of folding and faulting if the line is at right angles to strike; and it gives the necessary data for calculation if the line is oblique. What is needed, but not available, is a simple means to transform from hyperbolas to circles.

Some excellent examples of the interpretation of complex seismic sections, including sections with many hyperbolas, are figured and described by TUCKER & YORSTON (1973).

2.6 Tabulation of hyperbolas and their interpretation

To aid interpretation, the following tabulation has been drawn up.

2.6.1 The hyperbola has the same phase throughout.

The feature can be an anticline or a syncline: or the hyperbola may arise from a fault feature; i. e. the feature is not yet diagnostic.

2.6.1.1 The curvature of the hyperbola is greater than that of the NMO curve at this depth.

The feature is a syncline, with its centre of curvature below the surface, and the strike probably near the normal to the line of traverse.

2.6.1.2 The curvature of the hyperbola is equal to that of the NMO curve at this depth.

The feature is probably related to a fault or a sharp flexure whose centre of curvature is close to the folded horizon: The strike is normal to the traverse.
Exceptionally this feature is a syncline exhibiting two equal and opposite effects.

(a) Compression laterally as in 2.6.1.1 above.
(b) Stretching due to obliquity of the traverse ot the strike.

2.6.1.3 The curvature of the hyperbola is less than that of the NMO curve at this depth.

Generally this feature will be an anticline.
Exceptionally it will be associated with a fault oblique to the line of traverse, or to a syncline very oblique to it.

2.6.2 The hyperbola has different phase on the two sides.

The feature is a fault or a sharp flexure in which the dip is steeper than the distant ray paths on one limb of the structure.

2.6.3 The family of hyperbolas form a single alignment of vertices.

The feature will be anticline, a syncline or a fault, as determined by other evidence. Velocity and strike data may provide the basis for calculation. Correlation may provide other evidence.

2.6.3.1 The family of hyperbolas forms a double alignment of vertices.
The feature is a sharp double flexure.

It must be appreciated that this simple schematic cannot always be followed. Bad data will obscure the necessary evidence. Complex geology will do the same.

3 The 'black box' concept

3.1 Introduction

The system

The interpreter must regard the system which produces a seismic section as response to a succession of seismic inputs as a 'black box' – a single system. If the output from the black box were a perfect section the interpreter would not need to look further into it. However, the black box is far from perfection. Some understanding of how it works is necessary for the interpreter. The worse the section the more he must concern himself with it. It is in this chapter that noise and other effects which degrade the seismic section are considered.

The black box includes the earth, the recording equipment, a vast delay line and a processing unit. Input to this system is a series of seismic signals. Viewed in the time-domain they may be short pulses, as from an explosive source. They may be long swept frequency signals, as with Vibroseis [1]. At sea a number of different input devices are used. Most of them produce signals with an initial spike, but the spikes are followed by 'tails' which are the relics of bubble activity. Output from the system is the cross-section for the interpreter.

An ideal black box is one which carries out a very simple process. It convolves (see 3.3) the log of reflection coefficients (see 3.2) with the spike input: that is to say, it reads the log of reflection coefficients, and outputs it to the cross-section.

A real black box is far from perfect. The earth sub-system is much more complex, and its transfer function (3.4) is much more complicated than a simple convolution with the log of reflection coefficients. What the earth does, as expressed in its transfer function, is quite beyond control – it is a seismic expression of the nature of the earth. The recording and processing sub-systems have many parameters under the control of the geophysicist. By changing them he changes the transfer function of the sub-system, recording or processing. The purpose of changing the transfer functions in this way is to endeavour to undo all the work done by the earth transfer function except the convolution of the input signal with the log of reflection coefficients. If this can be achieved the output cross-section will present the required information and nothing else.

An additional function may be needed somewhere in the system if the input signal or 'signature' is not a spike. In such a case the reflections may need to have their shape altered to what they would have been if the input signal had been a spike. To do this it is necessary to have an element

[1] Trade mark of Continental Oil Company.

in the system which has a transfer function that accepts the signature as input and outputs a spike or something approaching it.

The sections that follow will first develop some of the terms used, and then examine the whole system in some detail.

The purpose in doing this is that the imperfections of the system may produce results on cross-section which simulate geological phenomena and mislead the interpreter unless he is aware of them. It is also the case that the interpreter, when he finds imperfections in his cross-sections may be required to direct or recommend the changes in the system to overcome the defects.

One of the themes running through this discussion is that of tuning. If all the operators in the black box are tuned to the operator of interest – which is the log of reflection coefficients – then it will appear loud and clear in the output. If the tuning is poor, then the log of reflection coefficients will be distorted or even rejected. A principal problem in the direction of seismic data gathering and processing is to identify which part of the system is de-tuned when the results are not what is required. It is unfortunately the case that many of the changes seen on cross-section could be the expression of changes in the log of reflection coefficients; or they could be the result of some change elsewhere in the earth or in the recording or processing sub-systems. The evidence will usually lead to one part of the system being the probable source of the change, but it will rarely be conclusive.

Good tuning improves the absolute amplitude of the signal of interest: and it is an important part of the quest for good signal-to-noise ratio. The second theme running through the discussion is the reduction or avoidance of noise, which is the other part of signal-to-noise improvement.

Most of the operations effected in the system are either convolution, or addition.

3.2 The log of reflection coefficients

The log of reflection coefficients is the seismic expression of the geological sequence. It is the theoretical limit to seismic resolution.

The reflection coefficient is defined as the ratio of the pressure amplitude of the reflected wave to that of the incident wave. Mathematically it is

$$R = \frac{\rho_2 V_2 - \rho_1 V_1}{\rho_2 V_2 + \rho_1 V_1}$$

where R is the reflection coefficient

ρ_2, ρ_1 are the densities of lower and upper formations

V_2, V_1 are the corresponding velocities of compressional waves.

It is seen that this can be re-stated in terms of the elastic properties of the rocks above and below in various ways.

The reflection coefficient is plotted against the reflection time to give the log.

The log is not tracked along a vertical into the earth, as if one were logging a drill hole. It is tracked along a line which is the locus of all points at which a ray from the observation point is normally incident on the reflectors. Expressed differently, this is the locus of all points of tangency of the advancing wavefront with the bedding planes.

Thus the line along which the log is measured is more like a bent pin than a vertical drill hole. At an angular unconformity there is a discontinuity in this line, but the rock properties are usually changing only slowly laterally, so that the discontinuity in the line does not have any material effect.

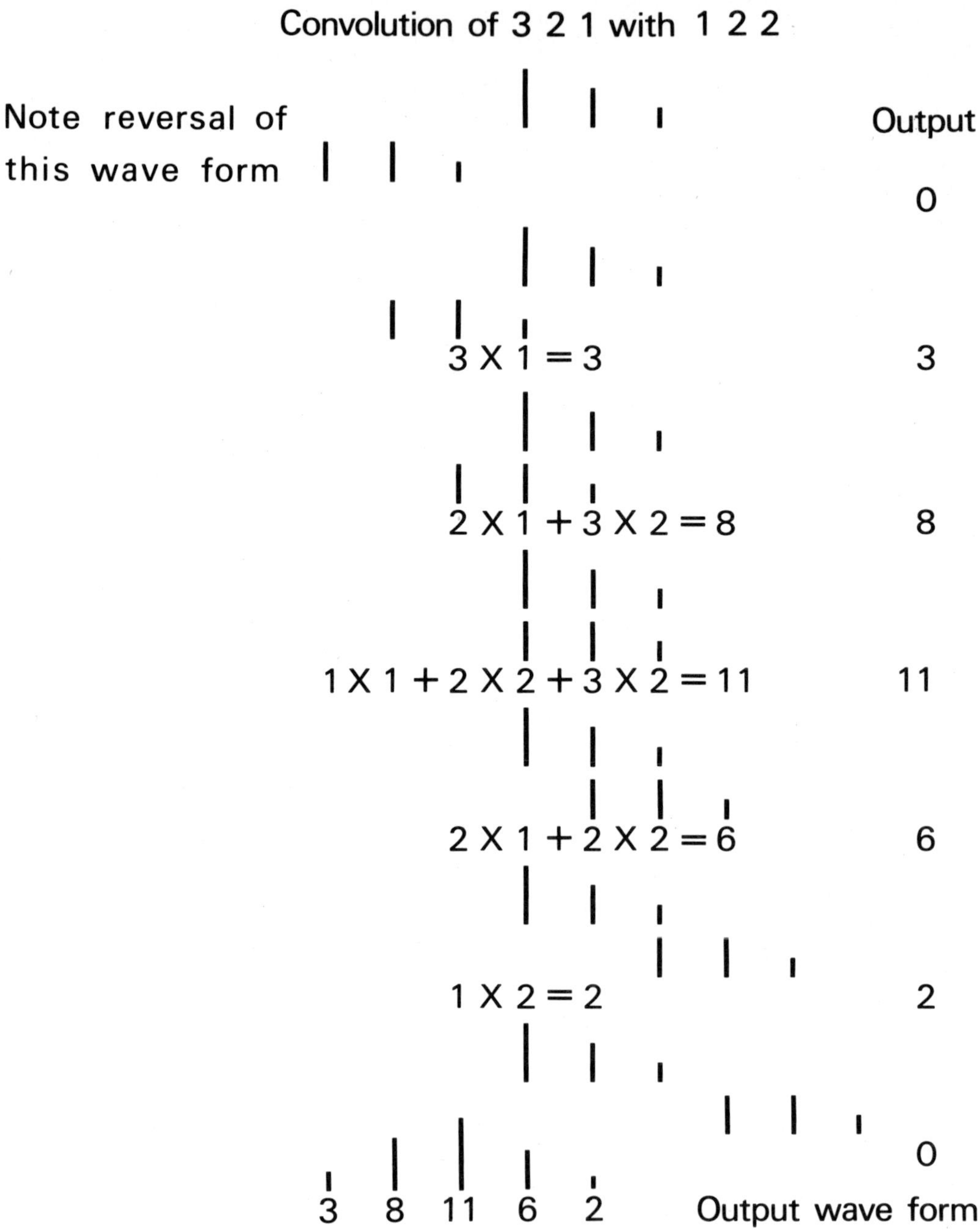

Fig. 24. A simple convolution example for those not familiar with the process.
It will be seen that the output of a convolution process must be longer than either of the inputs. If they are of length m and n the output is of length (m+n–1). Thus, every earth process – recording process – data process – must lengthen the seismic pulse if they can be regarded as convolutions. Some data processes set out to shorten the pulse. They can achieve it only by making some of the ordinates small, never by making them zero. In other words, they generate noise.
Note also that there are only 5 ordinates in the output and 6 in the input: one cannot discover anything about the input unless something – at least one ordinate – is previously known.

3.3 Convolution

Convolution is the process of sliding one sampled waveform over the time-reversed values of another, and out-putting the sum of the products of coincident samples. A simple case is illustrated in fig. 24.

A few generalizations can be made about convolution.

I. The output waveform is longer than either of the inputs, though, of course, some of the end values may be very small. An 'inverse operator' is one which reduces a waveform to a spike or nearly so.

II. Convolution with a unit spike makes no change: the spike 'reads' the other waveform.

III. Convolution of symmetrical waveforms yields a symmetrical output.

IV. Convolution of a long waveform (such as a log of reflection coefficients) with a short one (such as a seismic pulse) blurs the fine detail of the log without destroying its bolder features.

The mathematics of convolution is given in standard texts such as LEE (1960): and reference should be made to these.

It will be shown that the greater part of the processess throughout the 'black box' can be treated as successive convolutions of operators with the input signal, corrupted with noise of different constitution and intensity at the successive steps.

3.4 Transfer function

The transfer function of a system or sub-system is the function which operates on the system input to give the system output. In the seismic case the input signal is typically a spike from an explosive source. The only operator which is of value to the interpreter is the log of reflection coefficients: if it were the only one in the whole seismic system, then the output would be perfect for his use.

The design of a system, or rather the recording and processing parts of it, has as its purpose the determination of parameters, such that when the earth transfer functions other than the log of reflection coefficients are combined with the transfer functions of the recording and processing sub-systems, the output is a spike, or at least a fairly short pulse.

This brief discussion leads us to the concept of the input seismic signal being acted on by a series of transfer functions whose effects cascade as the signal makes its way through the complex system. The treatment is not an adequate discussion of the system. The purpose has rather been to establish some points of contact between the seismic process and signal and system theory. Some texts are quoted in the bibliography.

3.5 Earth, recording and processing contributions

This section reviews the processes taking place within the 'black box'.

In summary, the convolution processes in the black box can be regarded as operations performed on the log: convolution processes are commutative, so the order in which they are considered is not material. They are:

Convolution with the input signature. This signature is changed and broadened with travel time. An input spike (Dirac impulse) becomes increasingly stretched out as it travels through the earth. Any other signature can be regarded as the input waveform convolved with the input spike stretched out by the amount appropriate to the travel time.

Convolution with the operators expressing the action of superficial layers. There are one or two passages through these superficial layers. The operators which express their action are often severe. The superficial layers are often frequency selective to transmission, and the air-earth interface often produces a powerful multiple (or ghost) reflection.

In the recording part of the black box there are two principal operators which are of the same nature. The first results from the use of source patterns and detector patterns. Their general action can only be expressed in the wave-number or K domain, but for a specified angle of emergence this becomes convolution. Thus the operators will vary with the dip of the reflector and the distance of the detector.

The second operator is that due to the sampling and quantizing processes that together constitute digitizing of the continuous function.

The first can be a severe operator, the second has only minor effects in general.

Finally, within the processing part of the black box there are available many operators of great severity – bandpass, reject, or notch filters: deconvolution filters: pulse shaping filters.

The processes in the black box are associated with other additive processes. In summary, these are:

The addition of noise. This may be due to wind or other natural causes, in which case it may approach white noise. The noise may be man-made – trucks or airplanes, industrial machinery, pick-up from transmission lines or communications. In these cases the noise source may have a strongly peaked spectum.

Another purely additive phenomenon is the superposition of some other seismic modes, loosely grouped as ground-roll, and near surface refractions. These are generated only in the near surface rocks, and bear no relation to the rest of the sequence.

Within the black box, also, are data processing stages which are purely additive. For example, summing and stacking, whether carried out in a recording unit or a processing unit, are clearly additive. Some techniques of removing sea-bottom reverberation, and predictive deconvolution as used to remove section multiples, are essentially the addition of an out of phase estimate of the reverberant system after its first member.

In a separate class are some data processes such as the application of normal move-out corrections, which are designed to compensate for the effects of changing the source-detector geometry. This is a non-linear 'stretch' process applied to the data. It presents the log of reflection coefficients correctly, but the seismic pulse with which it is convolved is stretched in an irregularly time-variant fashion.

3.6 Compressional pulse transmission and reflection

Sedimentary basins make a layered earth, and all the phenomena of transmission and reflection of the seismic pulse must be considered in a layered model.

A downward travelling pulse loses amplitude. It loses it for several reasons.

Transmission through any medium tends to lose energy by absorption. This is frequency selective, diminishing the higher frequencies more than the low, and so widening the pulse.

Spherical spreading reduces the amplitude linearly with time where the wavefront is spherical – approximately in linear fashion, where it is not.

The advancing pulse loses amplitude by reflection at every interface through which it passes. All of these upgoing pulses are reflected downwards again at every interface, and so they add to the downgoing energy, but with a variable time delay. Those with only a short delay, the 'peg-leg' multiples of Anstey, follow within the length of the primary pulse. They build up its amplitude, change its shape, and generally impose a delay on the times of the peak and trough elements. Those with a longer delay build a waveform of separate elements later than the recognizable primary pulse. This is the 'tail' of the downgoing pulse.

Reflection of this pulse and its tail would be a simple reversal of the direction of travel. In practice a short section of the 'microstratigraphy' in the immediate neighbourhood of the interface – about the wavelength of the pulse – is involved. Each minor interface functions as a reflector, and the results are added with their time delays. This is a convolution process; and what is seen here is the reflection complex exercising its transfer function, or imprinting its character on the arriving primary pulse and wave train.

The upgoing pulse and train again passes through the layered earth above the reflector; again there is modification of the amplitude and shape of the primary pulse, and further development of the tail.

The description deals only with the first order multiples, with pulses that have been reversed in direction twice so that they continue on with a time delay. A system of nearly parallel reflectors is a reverberant model, and higher order multiple reflections will add still further to the tail of the primary reflection. At each multiple reflection amplitude is lost so that the high order multiples become weaker and are finally lost below the noise level.

3.7 Surface materials and their effects

Surface materials of the earth are characterized by a greater range of velocities and densities than is found in the deeper subsurface, and hence by much larger reflection coefficients. The stratification is much poorer. Included in the surface materials are many artifacts – roads and airfields, deeply ploughed fields, garbage tips and other waste areas.

This quick review shows how variable these surface materials can be over short distances. An Irish example was the change from Carboniferous Limestone outcrop to a peat bog of considerable depth within a few hundred feet.

Some impression of the range of near surface velocities can be gained from the following short table (see page 40).

It is difficult to generalize about such a range of materials. One can say that in many cases these surface layers contribute notably to the 'tails' of reflections, especially where they are well stratified. Where the configuration is irregular they are responsible for interference. Surface waves, and returning reflections are reflected or diffracted back on to the spread.

Another property of the low velocity, dry surface materials is that they are often poor transmitters of energy, rejecting the seismic spectrum wholly or partially.

Material	Compressional wave velocity m/s
Soil	100 to 200
Loose sand	200 to 2000
Loess	300 to 600
Embankments and fill	400
Loam	800 to 1800
Glacial till	430 to 1040
Glacial sand and gravel	380 to 500
Glacier ice	1000 to 4000
Concrete	3560
Steel	6150 to 6300

A surface input method used on land requires the pulse to travel twice through the superficial material; and this is generally more damaging to the transmission than a single travel through it as with explosives in a sufficiently deep shot hole.

3.8 Other modes of vibration

An ideal seismic source should produce only compressional waves in the input signal. This can be achieved at sea, and in other bodies of deep water. At the sea-bottom there is some partition of energy between compressional wave motion and other forms, but this seems rarely to cause problems in practice.

On land all seismic sources, and particularly surface sources, generate other types of vibration. Surface waves of various kinds are very powerful, often more than 100 times the amplitude of the reflection arrivals that they mask. Shear waves are generated also.

One must consider also refraction travel paths in the near surface rocks. These arrivals often obscure the earlier reflections.

All of these arrivals are functionally related to the geological configuration and rock parameters of the near surface. They are generated by the action of the appropriate transfer functions on the input signal. Theoretically, at least, one can imagine deriving the inverse operators which would reverse the action of those transfer functions, and recreate the near surface log from these arrivals.

In practice these arrivals are regarded as additives to the signal which is of interest, and all the conventional methods of treatment deal with them as such.

3.9 Noise

The earth motion which the geophysicist sets out to record consists of the wanted and unwanted parts of the earth response to the input signal.

The earth is a noisy place. There are many other seismic and acoustic sources in operation at any time. Those sources which are very close to the detectors produce signals which appear quite random – for example, grasshoppers on the geophones. Arrivals from more distant sources may be recognizable on many or all detectors. Distant sources that generate periodic waveforms such as ships, trains, aeroplanes, power stations and other industrial processes will add rather similar waveforms to all the data.

It is not only seismic and acoustic noise which concerns the exploration seismologist. Increasingly, electrical noise also corrupts the results. Sometimes this is of natural origin. The enormous amplitudes of electrical storms: natural electrochemical, seismo-electric, and other natural phenomena all contribute. Man-made sources: leaks from power circuits to earth or water, by-products from communications links, contribute locally to electrical noise.

The skills of seismic data gathering are directed to reducing the noise content in the recorded signal, and data processing to improving still further the signal-to-noise ratio.

3.10 The modification of signal in the recording process

One's first thought is that the function of the recording process is to record the earth motion as faithfully as possible. This does not represent current practice at all.

Recording is combined with two important data processing operations.

First is the use of patterns of input devices to generate the input signal, and of patterns of detectors to respond to the earth motion. These patterns are very powerful filters in the wave-number domain, accepting or attenuating signals on the basis of their wavelength as seen along the pattern. Patterns can be designed to obtain a good rejection of surface waves, or of horizontally travelling arrivals such as those from a ship's propeller.

Patterns are usually designed to have but little effect on reflections from horizontal beds recorded at short distance from the source. There is always some effect however, which can be expressed in the frequency domain for a given angle of emergence. The filter effect becomes more severe as the source-detector distance increases, or as the dip increases in the source-detector direction.

A second processing operation combined with the recording is bandpass filtering. A steep cut low pass filter, or alias filter ensures that the alias effect due to digitizing is greatly reduced. Typically for 4 ms sampling this requires a filter which is 80 dB down at 125 Hz: such a filter begins to attenuate at 62 Hz. A high pass filter which attenuates from 5 or 8 Hz, with much less steep cut-off, is combined with it to reduce the low frequency noise.

These two recording steps have a profound effect on the form of the signal.

Two other recording steps are of technical importance: both introduce noise: but neither has much impact on the interpreter. They are quantizing and multiplexing errors; GOLD & RADER (1969).

3.11 Reduction of reverberants in processing

It is seen that the earth and recording processes have taken the input signal and convolved it with the log of reflection coefficients. This is the required output.

It is seen, also, that the natural processes broaden and delay the primary seismic pulse, and add a tail to it. Natural and man-made phenomena add noise. Recording reduces some of the noise amplitudes, adds a little more.

The principal purpose of processing is to undo the effects of the last paragraph: to remove the tail and shorten the pulse: to reduce the noise in relation to the signal. The results have then to be presented in a range of amplitudes acceptable to the interpreter. One should add that the term 'shorten the pulse' also includes the effective shortening of the input waveform since, in practice, it is not a spike.

Methods for the removal of the 'tails' are well established, and they are described in the literature under the general heading of deconvolution. They are fairly effective: but the model used is too gross a simplification of the real earth, and the residuals from the least squares solution commonly employed are too large, for them to be fully effective. The process is further degraded by noise in the traces which leads to noisy estimates of the tail (or some function of it), and so to inverse operators which themselves contain an element derived from this noise.

The shortening of a pulse is well understood, but the practical problem is that of arriving at a good estimate of its form.

3.12 Improvement of signal-to-noise ratios in processing

An important method is that of combining a number of determinations of the value of the amplitude at each sample time. This is done by using a number of observations which have a common reflection point (on a vertical line mid-way between source and detector), but different source-detector distances. These are records obtained with a multiple coverage technique.

The first step is to correct the sample times for the differing geometry of these observations, back to the equivalent of a normal incidence record. This is the normal move-out correction, or NMO, itself a noise generator. Summation effectively reduces random noise and systematic or aligned noise which arrives at different times on the traces after NMO application. It is one of the more powerful processing steps.

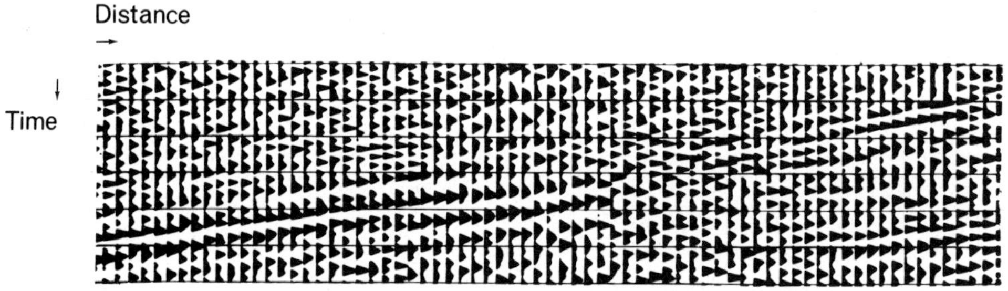

Fig. 25. The seismic response at the left-hand side is different from that at the right-hand side, and a direct seismic correlation is not possible. The two sides are separated by a noisy area which might be a fault.
The interpreter might be lead to speculate about rapid facies change, or various fault configurations which might produce this effect. In fact, in this old section, an abrupt change of filter was made at the noisy section, and it is probable that most of the difference between the two sides is due to filter change.

Bandpass or notch filtering can be used to improve signal-to-noise ratio where a clear frequency separation can be established between signal and noise. There is never a complete separation, since the seismic signal of interest has a wide spectrum. The decision which is taken is to sacrifice some of the seismic signal along with the noise so that the seismic continuity on what is left is more clearly visible.

A consideration for the interpreter is that long filter operators do something which is quite illogical, since they mix up observations from different parts of the geological sequence. The filtered Permian reflection may contain a contribution from the Tertiary.

The solution is to use short operators where they are effective, and to use long ones only where there is an improvement on balance.

Fig. 25 is an example of a small fault. Correlation across the fault has been spoilt by a change of filter.

3.13 Amplitude manipulation

Seismic signals as they arrive at the earth's surface decrease rapidly in amplitude with time, and the range is very large, often a million to one or more. The eye can appreciate this range, from 1 mm to 1 km, but it is clearly quite impractical to present the results in this range.

On standard cross-sections the eye does not appreciate anything much outside a 10 to 1 range in amplitude, from 0.1 mm to 1 mm. Amplitudes must be constrained within this range, either by automatic (data dependent) processes or purely functional changes.

The visual impression produced by the cross-section is dependent on the skill with which this last manipulation is carried out. The distinctive character of a reflection, and the resolution of fine detail can both be changed or lost by amplitude changes.

3.14 Summary of the effects of the 'black box'

This long digression on the 'black box' is a necessary introduction to those elements on the seismic section which are not of geological significance. The account has shown that these elements are of three kinds.

I. Alignments which are parallel to, and below a primary reflection alignment. These are either 'tails' of the reflection, or 'tails' on the input signature, which have not been adequately removed by processing. Where dipping beds are involved, parallelism may not be maintained. In marine surveys the reverberants involving sea-bottom and sea-surface may be very strong, and even sophisticated processing methods may leave significant relics.
There is a regularity in the spacing of successive multiples.

II. Alignments which have no obvious relationship to a primary reflection alignment. These may be residuals from surface waves, or from a distant noise source, which have survived the processing.

III. Random noise.

The problem set the interpreter is to trace his primary reflections across or through the interfering alignments of (I) and (II), and through the noise of (III) – or as far into it as he can go with confidence.

4 Interpretation on imperfect sections

4.1 Introduction

Chapter 3 has shown various earth processes adding unwanted alignments to the signals, and adding noise. Recording and processing sub-systems endeavour to provide inverse operators and signal-to-noise improvements so that the output is the log of reflection coefficients.

Such a design objective is never achieved completely. The assessment of what is signal is generally imperfect, and the techniques for removal of what is not signal are also imperfect. The final section provided for the interpreter will be degraded to some extent by residual unwanted alignments and noise, left behind after the processes have been applied. Sometimes the limit is set by the state of the art, sometimes by budget considerations. The quality of material submitted to the interpreter varies over a wide range: particularly in areas where exploration has been in progress for a number of years, and old seismic sections form part of the data.

The final sections will also show changes in the reflections as they are traced, changes which have no geological significance, but which are due to parameter changes in the transmission path, or in the recording or processing sub-systems.

4.2 Changes in the reflection

The discussion of chapter 3 has indicated that the shape of the seismic pulse is determined by the 'micro-stratigraphy' in the immediate neighbourhood of the geological boundary which generates the reflection. It is this shape which has geological significance.

The cascaded convolutions elsewhere in the 'black box' modify this shape. There is nothing in the final shape to tell the interpreter whether the changes in shape of the reflection pulse which he may find along the line of traverse are due to changes in the 'micro-stratigraphy', or in one of the other operators. There are, however, some observations which help to identify the reason for the change.

Most of these observations are concerned with the behaviour of other reflections on the traces where changes are seen on the reflection being followed or 'carried'.

I. If changes take place abruptly from one trace to the next on all reflections, the post-stack processing probably involves a parameter change.
II. If changes take place abruptly from one trace to the next over a limited time range, some time-variant process has probably a changed parameter.
III. If changes take place over the width of a spread, the source of the change may lie either in
    (a) Pre-stack process changes.
    (b) Abrupt changes in the near surface conditions.

IV. If changes take place gradually over all reflections the change is probably due to changing near surface conditions.

V. If changes take place gradually over part of the time scale, it is probably related to some change in the transmission characteristics of the earth above the time zone affected. This change in transmission characteristics will generally be expressed as a change in the character of the reflections in the upper part of the seqence.

VI. A change taking place on a single reflection, independently of changes on other reflections on the same traces, is almost certainly due to stratigraphic changes at the reflector: it has geological significance.

VII. Large amplitude changes may fall into the categories (I) to (VI). They may also result from the focussing and de-focussing effects of lens-shaped sedimentary features in the transmission path, or curved reflectors. In this case a number of reflectors will probably be affected by similar geometrical configurations.

VIII. Strong changes in interval velocity of a stratigraphic unit induce large amplitude changes in two reflectors – at the upper and lower interfaces.

The change in shape may be in several different directions. It may simply be a shape change which preserves the classical trough-peak-trough or peak-trough-peak configuration. It may be an increase in amplitude.

Neither of these changes creates much of an interpretation problem; the continuity is not interrupted, and the reflection can be followed through. The interpreter will be aware that the change in shape may be shifting the maximum or minimum amplitude slightly with respect to the geological boundary which generates the reflection. In case (VIII) above, he will be aware that the shape of the lower reflection will be distorted from its real earth shape by the velocity change, and deeper horizons may show the same distortion.

A more troublesome change is the loss of amplitude of the reflection, to the point where its identity and alignment are lost below the noise level. This can be the result of the cascaded operators changing in such a way that they no longer pass much signal from that part of the log of reflection coefficients – the reflection complex is 'tuned out' by the changing operators.

In this circumstance there is nothing the interpreter can do, since he has no continuity to observe. If this part of the geological sequence is of importance he should map higher and lower reflections.

It will usually be the case that re-processing, directed specifically to recovering the continuity on the required reflection, will succeed. The only exception would be where the de-tuning occurs in the earth transmission path, and this case has not been demonstrated.

Sometimes the continuity is lost over the crest of a strongly folded anticline. This may be due to de-focussing: but it may equally be due to the special sedimentary or structural developments sometimes found in such crestal areas. No generalizations can be made.

The most troublesome change is the splitting of a reflection into two without geological reason. There is no logical basis for choosing one element rather than the other. Whichever route is followed it introduces a small time change which is not an expression of any structural change.

An empirical rule is to follow the earlier element in such a split reflection. This is consistent with the practice of using the earliest element of the reflection for mapping. If a reflection does split in this way, the pulse is likely to be longer after splitting. The rule must be exercised with care. Any structurally high area associated with a split must be re-examined with special attention. It is only completely acceptable if the lower reflection at the split would show the same structural high of lower amplitude.

Again, in critical cases, it may be necessary to re-process the data to remove this feature.

One might be tempted to argue that recording and processing practices should use parameters which are constant both with time and distance. It is, of course, the problems set by the real earth which require parameter change.

In strongly folded regions the use of time-variant filters may cause troublesome changes from one structural level to another, and filters may have to be made invariant to overcome it.

4.3 Interfering alignments

Where primary reflections are follwed by residual tails generated in a near surface reflection complex which is changing but slowly along the traverse the primary alignments are followed by near-parallel multiple alignments. The pattern of multiples is much the same after each reflection.

More distant elements of the tail are often multiples developed between two primary reflections. If these have an angle α between them, the angle between the lower reflection and the first multiple will be α, between the lower reflection and second multiple will be 2α, and so on. The most powerful multiples following this pattern are those reverberating between the surface of the sea and a hard sea-bottom. Special techniques are required to remove these 'trapped reverberations'. They are rarely completely successful.

There are a number of surface or shallow modes of vibration which arrive on the traces of the field recording at about the same time on many of the records along a line of traverse. Refractions and ground-roll are the most important. In some cases residuals from these arrivals will stack up to give nearly horizontal alignments.

Finally, there are some noise sources which produce periodic waveforms, (3.9). If records are taken at precise time intervals, the later processing can build up the interference into a persistent alignment which can be hyperbolic in form or curved in other ways if the source is moving.

Marine surveys are subject to this interference from ships. Some Vibroseis surveys on land can also experience it.

Wherever angular unconformities exist in the subsurface, there will be parts of the seismic section where the slopes of primary reflections and other alignments differ. Provided that the primary reflections are not too weak in amplitude in comparison with the interfering alignments, the geometrical relationships will establish the primary reflections and they can be traced through the interference pattern.

Where the angular differences are slight, the primary reflections and the interfering alignments are not at first recognized as separate elements. Instead, the dominant alignment is recognized, and it is seen that its amplitude becomes larger and smaller as the lower amplitude alignment reinforces or diminishes the dominant one. If the dominant alignment can be established as a primary reflection, say outside the region of interference, then continuity can be carried. If it is the interfering one which is dominant, then the primary reflection is represented only as the high amplitude or low amplitude parts of the interference pattern. These cannot be connected up with confidence unless there is other good supporting evidence: it is too easy to alias, to join up the wrong maxima or minima.

The high amplitude or low amplitude criteria could be used over a short distance to link two 'islands of confidence', but not otherwise.

Where the sequence is horizontal or uniformly dipping there may be no slope differences to aid discrimination. On such a seismic section there is nothing but the time difference relationships which can identify multiple possibilities: it is not conclusive, because one can never exclude the possibility of a primary reflection existing at about the time of the calculated multiple.

Here again it is advantageous to identify the primaries outside the horizontal or uniform dip area of the cross-section, and to trace them into the uniform dip area.

4.4 Noise

There is little that the interpreter can do about noise on the seismic sections. In most sections it sets the limit to the depth at which he can work.

About the only aid in noisy sections is to hold the paper a little below eye-level, and look along the suspected alignment. This seems to act as a 'visual stack', and to improve the signal-to-noise ratio.

It is a useful exercise for the interpreter to do some work with sections made with a low fold of stacking, and then to compare his interpretation with a full-fold stacked section. It is often satisfying to see that weak alignments on the low fold stack are confirmed.

5 Some sedimentary features on seismic sections

5.1 Introduction

Previous chapters have dealt with those earth features which can be considered as part of a 'layered earth'. The horizontally layered earth is a starting point for most seismic modelling. The model is changed in a simple way by folding and faulting, and its seismic expression blurred by processes which have been described.

In this chapter the features described introduce elements into the earth model which are far from the horizontally layered idea. They correspond with palaeogeographic environments which are changing over rather short distances: and the sediments, and their seismic expression, are changing correspondingly.

Interpretation and description of these features must be in sedimentary terms as well as structural terms. Structurally they may be insignificant.

5.2 Deltas

Deltas are highly complex environments, but only a few of the significant features can be seen on seismic section.

A simple classification of deltaic sedimentation is into bottom-set, fore-set and top-set sediments. Fig. 26 illustrates a typical fore-set bedding. The reflectors dip rather steeply between limiting upper and lower reflections which are nearly horizontal. A fore-set reflection traced upwards is seen to pass into a near-horizontal top-set reflection: traced downwards it passes into a near-horizontal bottom-set reflection. The shapes of the reflections thus illustrate the pro-grading of a delta: the way in which it builds out over pre-existing shelf sediments with the 'tip-heap' configuration of the fore-set beds. It illustrates also the flat-lying, swampy sediments of the top-set, where the delta has built up to about sea-level, a little above or a little below.

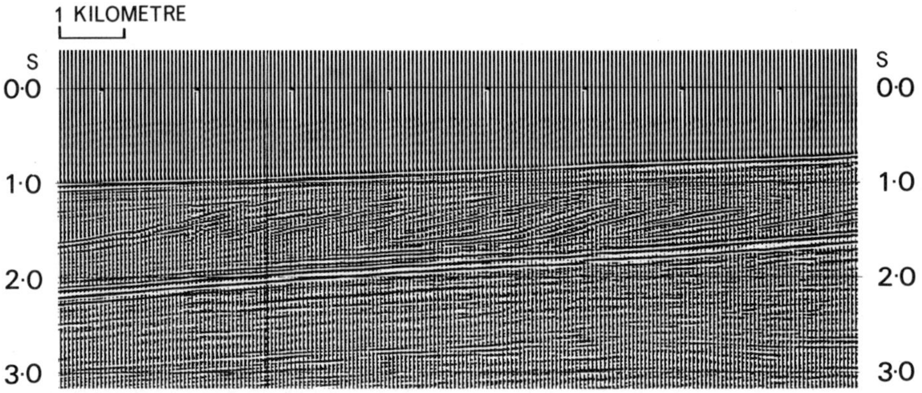

Fig. 26. Deltaic sediments, showing the classical relationship of top-set, fore-set and bottom-set beds. Individual reflections can be traced through all three groups.
Note at the right-hand side an extensive washout which cuts deep into the fore-set bedding, but is itself overlain by top-set beds which in their turn pass into fore-set and bottom-set beds as they are traced to the left.
This section preserves the record of a typical interlude in the history of a delta. A river changed its course and eroded a channel in previously laid down deltaic sediments close to the edge of the delta, with typical 'tip-heap' sedimentation of fore-set beds a little further out. The channel was silted up and abandoned, and a uniform marshy surface restored, by which time the front of the delta had advanced a little further to the left over the bottom-set beds.

At the foot of the fore-set reflections there are often small areas of disturbance. These may be due to small-scale slumping. More often they seem to represent disturbance of the bottom-set beds by currents down or along the face of the pro-grading front of the delta.

Another feature of deltas is that the distributaries of the river running through the delta frequently change their courses. They are often constrained by levees: they build up the levels of the river bottom by dropping sediment, until the situation becomes unstable. The distributary breaks through the levee and establishes a new course.

Seismically, this appears as channels cut through the upper parts of the deltaic sediments – the top-set and fore-set beds. These channels are infilled after they have been abandoned, often, but not always, by rather featureless sediments showing poor reflections.

Fig. 26 illustrates typical features of deltaic sedimentation, including channeling, overlain by top-set beds, on the right-hand side.

Many features of deltas are reproduced on the edge of the continental shelf, where pro-grading fore-set beds, sand and mud, build out over the edge of the shelf. A special feature of this environment is that the angle of rest of the fore-set beds is often higher than that of the underlying continental slope sediments. The mass of fore-set beds becomes unstable, and slumps down the face of the continental slope.

In the Gulf of Mexico, these great masses of unsorted material contribute to the infilling of the recently formed peripheral sinks surrounding salt pillows and domes. They appear on seismic sections as thick lenticular masses, largely free of reflections. See 5.4 below, and LEHNER (1969).

5.3 Reefs

The classification of reefs and reef materials by geologists and others is complicated. The processes of reef-building and the associated sedimentation, and the later changes in reef rocks and their associates can take many forms. Sometimes the terms are used rather loosely. Seismically one is concerned only with reflections and the shapes that they show, and this leads to use of older terms, based on form, for descriptive purposes.

A primary distinction is between bioherms and biostromes. A bioherm is a mound-like structure built by sessile benthonic organisms, and enclosed in rocks of different type. Such a reef exhibits the lithological differences necessary to generate reflections at its boundary, and a sufficiently distinctive shape for it to be recognized. The term is rather broad in its usage, and although it generally refers to wave-resistant structures, the popular idea of a reef, it has been stretched to include wave-built structures. These last are wave-built banks or mounds of the debris of sessile animals and other calcareous remains.

A biostrome is a fairly regular bed, composed of calcareous material, some remaining in place, some having been wave transported. Some biostromes represent reef complexes: widespread calcareous developments which had biostromes developed from them by planation by erosion. Advance and regression of the sea may so reduce a bioherm that only a near-uniform bed remains, as a biostrome.

The seismic distinction is that bioherms can be recognized and detailed on seismic sections, biostromes have no characteristic form. Biostrome recognition depends on their identification in wells, and the tracing of continuity, together with continuous interval velocity studies, away from the control point.

Typical seismic expression of a reef, or more exactly a wave-resistant bioherm, shows the following features.

I. A lenticular shape, with the upper face convex upwards, the lower essentially plane.
II. A steep fore-reef slope, sometimes steep enough to develop hyperbolas at the force-reef side of the reef core.
III. A more gentle back-reef slope.
IV. A difference between fore-reef and back-reef sedimentation, so that the seismic sequences do not correlate in detail.
V. Draping of sediments over the reef: the convex upward form of the upper reef surface persists, with decreasing relief, into higher beds.
VI. The detailed structure of a typical reef is a reef core, massive and unbedded, with fairly steep dips away from the core. Evidence of steep, persistent reflections can often be seen.

VII. All of the complex described under VI, reef core and reef flanks, is likely to be limestone with a wide range of biological content and detailed structure, but all with a high velocity. The effect on seismic section is to decrease reflection times of deeper reflections as they pass under it. This is the so-called 'velocity pull-up', so that the under-surface of the reef, described as 'essentially plane' in I, is generally slightly convex upwards on seismic section.

The original animals had extensive cavities in their hard parts, and so the original reef had high porosity. This is retained in young reefs. Older reefs become cemented up and lose their porosity. Still later, other changes such as dolomitization may induce a secondary porosity. The resultant high porosity and low velocity are less common than the high velocities described above: they may establish the commercial interest of reefs.

This generalized description applies to a wide range of reef forms as seen in seismic section. Barrier reefs, hinge-line reef developments, and isolated atolls, all show the basic asymmetry described.

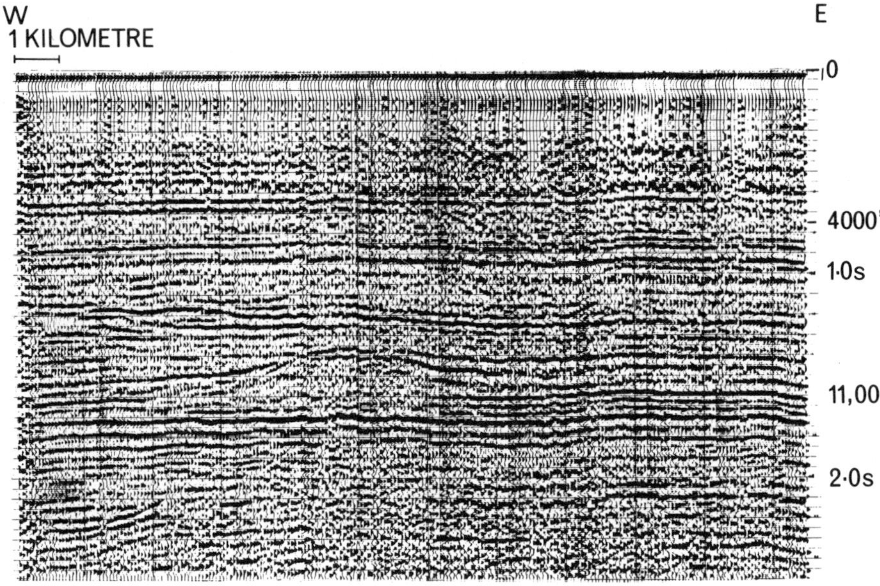

Fig. 27. A coral reef with its floor at about 1.8 s and its crest at 1.35 s. Note:
(1) The small hyperbola developing on the left-hand side which marks the steep fore-reef face of the reef, and the unbroken seismic expression of the gentler back-reef face.
(2) The difference in the seismic response of fore-reef and back-reef sediments.
(3) The "velocity pull-up", or apparent gentle anticline in the floor of the reef which is an expression of the higher velocity of the reef than of the off-reef material, and consequently lowered reflection time for sub-reef reflections. This "velocity pull-up" feature of the seismic response has no corresponding structural expression in the real earth.
(4) Above the reef there is a gentle arching of the sediments due to draping over the reef. It can be detected as high as 0.65 s. This is a real earth structural feature.
(5) Near the bottom left corner is a steep dip indication which suggests a possible earlier reef horizon. The data are not adequate for a confident interpretation.
(6) At 1.2 s, left of the reef, is a small pod-shaped mass bounded by reflections. This may be a biostrome, a bank of calcareous debris from organisms, wholly or largely wave transported to its present position.

Seismic sections of reefs may show a symmetrical development. This may arise if the section does not include fore-reef and back-reef regions – if it is at right angles to the asymmetric section. It may arise also if the reef is an isolated one, more or less circular in outline, with no asymmetry.

A feature of many recent reefs is the highly irregular upper surface, often extending into pinnacles, which the living organisms create. On seismic section these may appear as isolated hyperbolas, looking like umbrellas on the sea-floor, or as large families of hyperbolas. These irregular, fragile structures do not usually survive later advance and regression of the sea, and fossil reefs, seen deeper in the seismic sections, are usually much smoother.

Smaller, rather featureless lenses appear in the nearby sediments. These may be small satellite reefs whose characteristic features cannot be resolved seismically, or they may be wave-built mounds of debris.

Coral growth and that of other sessile organisms is largely controlled by water depth and food supply. Barrier reefs offshore, and hinge-line reefs occupy areas where the required environment may persist, continuously or intermittently, over long periods of time. The reef complexes then become very elaborate: though they are made up of the same elements in the seismic sections.

Fig. 27 is a well-expressed reef in seismic section, illustrating most of the features described. It will be appreciated that good quality sections are needed to resolve the diagnostic details.

5.4 Evaporites

There is little which is distinctive seismically about evaporites, as originally laid down. Since they are laid down in restricted basins, overlap or off-lap effects are commonly seen: but these are not diagnostic observations.

The features of seismic section which relate to evaporite basins appear where the process has gone far enough to produce saline deposits, and these, in their turn, have undergone halokinesis, the term applied to salt movement. Seismically, these salt movements produce salt shapes which are distinctive. The salt bodies affect profoundly the course of later sedimentation and its seismic expression. Because of the high velocity of salt, the salt shapes also impress their effects on the underlying rocks.

Many stages can be recognized in the development of salt structures and related phenomena as they are seen on seismic section. Good geological descriptions are given by TRUSHEIM and others. Not all of the stages described below can be recognized for every salt occurrence: in many cases the halokinesis stops short of the ultimate development. In others the developments in the subsurface, or the seismic expression of them, are incomplete.

The starting point for this description is a regularly bedded evaporite sequence including some saline members. A typical sequence of developments is traced.

I. Formation of salt pillows.

Salt is a mobile rock on the geological time scale. Surveys in salt mines have found movements of a few centimetres a year. It has been estimated that under natural conditions the movement is of the order of 1 millimetre per year.

Two different mechanisms have been proposed for the initiation of the salt movement. One view would consider that differential pressures on the salt layer would arise from effects in the overlying rocks. Differences in thickness or density in parts of the overburden would have this result, and would initiate a movement of salt from the high pressure area to the low.

The alternative view would consider that differential pressures would originate from effects in the underlying rocks, for example, by renewal of movement along a fault, or renewal of local folding movements. The high pressure area would then move upwards even against a uniform load of overburden.

The evidence from seismic sections is not conclusive: one can observe differences in overburden, and pre-salt structures. It is possible that both mechanisms may be operative. Whatever the origin, the salt pillow is a thicker development on the salt bed: and the surrounding salt is thinner – sometimes down to zero thickness. Typically the salt upper surface is smooth and continuous.

In plan the pillows are usually circular. In some cases, especially where the salt is very thick, they are elongate.

II. Migration of salt into salt pillows will arch up the overlying sediments into anticlinal or domal features. Between them are structurally lower features. Sedimentary processes will now begin to fill up these small basins. Above the smooth upper surface of the salt a certain thickness of sediments will show reflections paralleling this surface. Above this level the basin infill will become progressively less concave upwards, i. e. the sediments are thicker in the middle. At still higher levels the reflections resume the near-parallel realtionship of a regularly bedded series.

As one moves up the sequence of reflections above the salt, the two relationships described are important geological dates. The departure from parallelism of the reflections marks the first movement of the salt, and demonstrates the overburden thickness under which it took place. The resumption of parallelism dates the end of salt movement (or at least that phase of it), and the completion of infilling of the basin.

In a large salt basin the initial and final salt movements are not necessarily contemporaneous for all the pillows. A consequence is that a reflection from the upper or lower sequence of parallel reflections associated with one pillow may lead into the thickening trough sediments associated with another. Moreover, the sedimentary environment, and hence the stratigraphic sequence of the trough may differ from that of the shallow water areas so much that the pattern of reflections changes, and continuity or correlation may be lost. Even on the two sides of the same pillow the reflections may indicate this differing development.

A trough developed in this relationship to a salt pillow is a 'primary peripheral sink'. It will take on an approximately annular or crescentic form around a circular pillow: it will be two parallel troughs bordering a linear salt pillow.

III. The velocity in salt is high, reaching nearly 5000 m/s, while the velocity in the infilling of the peripheral sink may be a little lower than that in the sediments overlying the pillow. The combination of these two effects on deep reflections from below the salt horizon is to decrease reflection times under the pillow and increase those under the sink. The result is a 'velocity pull-up' rather like that described under a reef.

IV. In a second stage of salt movement the salt is largely expelled from the trough areas, and concentrated in salt domes which have typically much steeper sides than the pillows. There is usually no extensive fracturing of the enclosing rocks at this stage.

V. Removal of the salt from below the edges of the primary peripheral sink brings the floor of this sink back to a nearly horizontal position. Since the infill is thick in the middle, and thin at the margins, the originally flat surface of the infilled basin is now convex upwards. This is the saddle-back structure, a fold feature characterized by a *decreasing* amplitude with depth, a feature which can be seen on seismic section.

VI. Between the saddle-back and the salt dome is left a depression, the secondary peripheral sink. It is much smaller in diameter than the primary one: but most of the features described in II. for the primary sink will be found in the secondary one on a smaller scale.

The secondary sinks have sharper curvature than the primary ones. In depth the centre of curvature is usually below the surface, and this sink shows the convex-upward seismic expression of a syncline as described in 2.3.

VII. The velocity pull-up described under III. is much more severe with this more advanced stage of salt migration. Moreover the strong focussing and defocussing effects of the geometrical shapes of salt and peripheral sinks create problems in determination of normal move-out (NMO), and in recovering a signal of sufficient amplitude, for horizons below the salt.

VIII. The crestal regions of salt domes as seen in seismic section are often complex. Continuity is often interrupted, and hyperbolas are developed. Sometimes this is simply a consequence of the sharp curvature of the crest and its steep flank dips. In other cases, especially where the salt structure is shallow, the confusion appears to be the result of solution collapse in the crestal part of the dome. LOHMANN has described solution structures. Complex faulting in the crestal regions of strong anticlines can also produce a complex seismic expression.

Such areas of complexity are difficult to interpret, and one cannot always establish the origin of the seismic features.

IX. Within the salt beds there may be beds of some other rock which will generate reflections from their salt contacts, e. g. anhydrite, dolomite, black shale. These beds do not have the high mobility of salt. When the salt moves into pillows or domes these other materials are carried along, and must adjust their shapes to the new configuration. They appear to break up into faulted 'rafts' or to be thrown into a series of sharp folds. The seismic expression is an irregular chain of hyperbolas, closely spaced, with their vertices at about the same time.

X. The next stage in salt movement produces a piercement structure: a plug of salt which breaks through the overlying sediments. The truncated ends of these beds generate hyperbolas which obscure the details of upturning against the salt.

XI. This late stage of salt movement removes salt from under the inner edge of the *secondary* peripheral sink, and produces small, convex-upward folds. The new, tertiary sinks so formed are very narrow and sharply folded. They are rarely defined seismically, but produce more hyperbolas. It is generally difficult to distinguish between truncation and tertiary sink hyperbolas.

XII. The final stage of salt movement is its extrusion at surface. This surface salt has no seismic expression; and in fact it is dissolved by water in all but the most arid environments. In some marine surveys the approach to the surface can be recognized by a slight rise on the sea-bottom, and a chart notation of 'boulders' which represent the insoluble rubble brought up by the salt.

6 Seismic terrains

6.1 Introduction

Experience shows that in any survey which is being interpreted, the nature of the problems to be solved is much the same throughout the area. The problems may be quite different elsewhere. One is able to group similar surveys into 'terrains', characterized by the same class of problem. These are sometimes, but not always, the same divisions as between geological provinces.

In some areas the upper and lower parts of the sections offer quite different problems, or terrains. One tends then to classify the survey on the basis of its character in the zones of commercial interest.

6.2 Nearly horizontal beds

In areas with nearly horizontal bedding, the commercial interest is likely to be in structures of low relief and very gentle slope, or in stratigraphic traps that is pinch-outs up-dip due to gentle convergence of beds.

The interpreter's need in such terrain is precision. Near surface corrections must not be in error. The seismic pulse must be as sharp as possible.

Multiple reflections are the greatest source of potential mistakes. The culminations of gentle structures can change in position with depth as a result of slight thickness changes. Unrecognized multiples will mislead the interpreter into continuing downwards the same culmination. Similarly, slight convergences can be created or obscured by the presence of multiples.

6.3 Strong fold-structure

Surveys of this group are perhaps the most interesting to interpret, and probably the ones which can be interpreted with the highest level of confidence.

Folds, associated faults and unconformities are all bold features, and if the horizons are readily traced the interpretation is often straight-forward.

The special problems of such a terrain are those of correlation. Synclinal relics or down-faulted blocks can be widely separated below the present erosion surface or below some major unconformity. Cores of anticlines or up-faulted blocks may contain reflections which cannot be traced through the deep areas where the reflections are lost below noise level.

It will be seen that correlation of these relics, often widely separated, may be difficult because of facies changes or simply because the relics are not stratigraphic equivalents.

6.4 'Hyperbolas' terrain

The seismic sections from some surveys are dominated by large numbers of hyperbolas. The methods of interpretation are broadly those of 2.5. This terrain is, geologically, of very mixed origins. The structure can be sharp folding at very shallow depth, moderate at intermediate depths, or almost any departure from the plane at great depths.

The structure may be dominated by faulting, of any pattern.

Sharp, irregular features of sufficient size, i. e. comparable with a wavelength, can function as point sources. These may be the irregularities of coral reefs. They may be the buried scarps and other topographic features of hard rocks buried at an unconformity. They may even be great boulders or erratics in sediments of the largest grade size.

Finally, promontories on an irregular coast-line, transmitting devices on land or on the sea-bottom, reflections from the hulls of other ships, their engines or other sources of industrial noise can produce hyperbolas on seismic section.

It is probably in the interpretation of these 'hyperbola' sections that the interpreter works with least confidence. Interpretation depends on a precise knowledge of strike and of velocity, and these parameters are often hard to establish. As will be discussed later, migration can make the visual assessment easier; but the correctness of the migration process is equally dependent on the same parameters.

6.5 Stratigraphic trap terrain

The search for stratigraphic traps is largely a search for converging reflections, and recognition of the points at which particular potential reservoir rocks pinch out, or can no longer be traced by their upper and lower reflections. The exercise is one which requires careful tracing of continuity and its check by correlation. Recognition and rejection of residual multiples is of the highest importance: these residuals can simulate the convergence of primaries and the pinching out of stratigraphic units.

6.6 Sedimentary features terrain

In such an area the interpreter is concentrating on the special shapes which reflections take around the principal sedimentary features, and on the fine detail associated with them. Structure must still be determined as accurately as can be, since hydrocarbon migration and accumulation are controlled by the shapes and permeability and porosity distribution which results from sedimentary and structural processes.

Seismic recognition of sedimentary features is especially dependent on having good sections. Noise can readily obscure the important diagnostic details.

6.7 Noisy terrain

Occasionally the interpreter is required to work with reflections which are dominated by noise. This may be 'amplitude noise' – noise signals added to those of interest and obscuring or obliterating them. The noise may be 'time noise'. This term is used where the recorded times of reflections are scattered about their true positions. The continuity is no longer a clear alignment of corresponding elements, but a vague band of higher amplitudes than the surrounding background.

With such sections the interpreter must do his best. The bolder reflections may be recognizable, and the course approximately lined in. Details of structure and sedimentation are lost, and the resulting interpretation is smoothed and generalized.

One of the report items in dealing with such a survey is to give the points of interest, as far as they can be determined, and the steps which should be taken to improve the quality of the sections at those points. It is important, also, to indicate the 'measure of confidence' which the interpreter has in different parts of the work.

7 Presenting the results of interpretation

7.1 Introduction

Most of the discussion so far has been two-dimensional: the interpretation of what is seen on the seismic section. The earth is, of course, in three dimensions, and no geological problem is solved until the answer is in the solid.

Seismic observation is curious in that it is made in fine detail along lines of traverse, with nothing between. In consequence, interpolation between traverses plays an important part in interpretation and presentation of the results. Usually the presentation is reduced to two dimensions again, as contour maps, which gives even greater prominence to the interpolation.

Hitherto the vertical dimension has been in reflection time. In the presentation there must be a discussion of the depths of the horizons mapped, either as a series of maps contoured in depth, or an indication of the possible nature of the change from the time maps if good velocity determinations are not available.

7.2 Timing reflections

The first task is the routine one of reading the reflection times from the cross-sections of the survey. The interval at which this is done will depend on the scale of the final map which is to be produced; the choice is governed by the wish to have a suitable density of control. About one observation per centimetre of traverse on the map is a reasonable density.

The time zero is shown on the cross-section, and the time scale is shown on the margin. Usually time lines at short intervals are drawn across the section. This guidance, and a scale to the cross-section time scale is all that is needed to time the reflections.

The element to be timed is either the line drawn by the interpreter in tracing the horizon, or the peak or trough which he is following. Ideally these are the same: but in a noisy section the interpreter may wish to smooth the peak or trough times from trace to trace visually, and does this by drawing a smooth line. On the other hand an interpreter who is in a hurry may draw a line following roughly the course of a very good reflection, with the intention that peak or trough times are used for mapping. The intention must be made clear, especially if this phase of the work is delegated.

Reflection time readings are listed or entered directly on base-maps for each horizon. They must always be checked.

7.3 Structure maps in time – isochrons

A first assessment of the structure of a seismic horizon which has been timed is made by taking a plot showing the lines of traverse and the shot-point or trace numbers for which the times have been read. The times are entered against the corresponding data point, and these data are contoured by lines of equal time, or isochrons.

Such a map serves to show the fold axes and faults, and some measure of the flank dips, provided that the slopes involved are not too gentle. The approximation of time map to structure becomes less and less secure with increasing depth. For example, a flank of 10 ms on a structure at 0.2 s, average velocity 2000 m/s will not disappear on conversion to depth unless the velocity decreases, in the apparent downdip direction, by 4.8% over the 10 ms flank. On the other hand, a 10 ms flank on a 2 s reflection, average velocity 3000 m/s, will disappear with an average velocity decrease down the 10 ms flank of only 0.5%.

There is one factor which will always increase the velocity in the downdip direction, that is the effect of increasing load on clays and shales. With shallow dips, however, this can be more than offset by facies change, or by lenticular developments of salt, limestone, basalt, and other high velocity or low velocity rocks.

It is in the contouring of a time-map, the drawing of isochrons, that most interpreters are introduced to the concept of interpolation. At this level one does not wish to analyse the problem too deeply, but rather to derive some working rules.

The most secure interpolation is that which involves nearly linear features or functions. In the seismic case these are fold axes, and faults which are nearly vertical. The intersections of these features with the lines of traverse can be recognized and marked. Since the interpreter has been building in his mind a three-dimensional picture of what the subsurface is like, he will be able to express this form in the way in which he joins up these indications of folds and faults across the empty grid spaces.

It should be stressed that this 'joining up' is an important interpretation phase. There is nothing in the data which will reveal by an analytical process that the faults seen on two cross-sections are, in fact, the same fault. The test to be applied here is that of 'geological plausibility'. One looks for similarities in the direction and amount of throw, and in the structures associated with the fault. One looks for the dates of the initiation and end of the fault movement. Initiation of faulting is dated by tracing up the fault plane until one finds the top of the zone of constant throw. The shallowest reflection to show this maximum throw dates the initiation of faulting. The shallowest reflection to be displaced marks the end of the principal fault movement, though gentle renewal movements may induce flexures at still higher levels.

'Joining up' of fold axis indications is usually more obvious. If the survey has been well planned, the grid lines will approximate to strike and full dip sections: the first has much less structural relief than the second, so that the interpreter has already a well established concept of the strike which will guide him in lining up the fold axes.

Even with fold axes there are difficulties. Figs. 28 A and 28 B for example, illustrate a frequently recurring situation. A grid square has structurally high observations on two opposite sides: structurally low observations on the remaining two sides. This can be an anticline, a syncline or a saddle. Only the test of geological plausibility, or the introduction of bias analytically, will help the

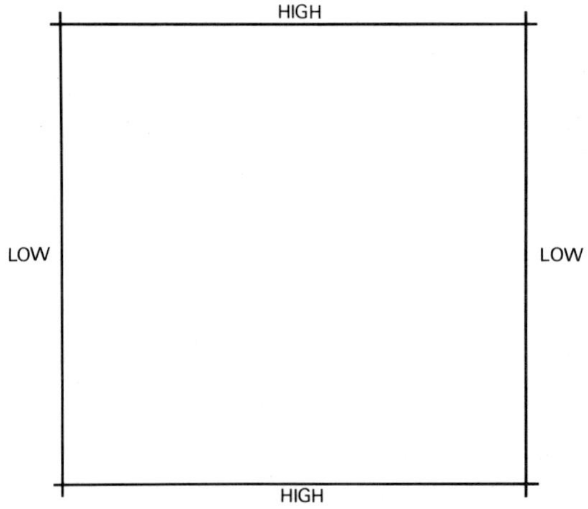

Fig. 28 A. The interpreter's dilemma. Has he a north-south anticline or an east-west syncline? Or has he, as most machine contouring programmes would give him, a symmetrical saddle centred in the grid square?
Usually the interpreter will accept a strike direction clearly established in the rest of the survey area, or by geological evidence. It is important to keep in mind, and to report, that the syncline or anticline has *not* been established by the seismic method, which has contributed nothing in this grid square. All one can say is that there are not, as yet, any seismic observations which are at variance with preconceived ideas.
A further possibility is that there are four isolated structures, two highs and two lows, of much smaller size than the grid square.
Whatever solution is adopted, it must be kept in mind that the contours in this grid square are controlled entirely by the interpolation procedure or formula adopted, not at all by the structure.
If the uncertainties raised by this illustration are sufficiently important, a better solution will be found by running further traverses. These new traverses can be designed to give denser coverage all over, so that if any doubts of the same kind remain they affect only a small grid square which is not significant. Alternatively, the new traverses can be designed to prove or disprove a particular hypothesis: this can often be done with less than complete dense coverage.

interpreter. In cases where resolution of this grid square is important, the interpreter may need to call for more data.

A different type of problem exists where the structurally high and low areas are not fold structures with a linear configuration, but are more nearly circular and discontinuous. Such developments arise in areas of salt tectonics, some areas of reefs, in volcanic areas, and in some areas of flat-lying sediments where the influences shaping the structure seem to be little more than differential compaction. Under these conditions there is no 'joining up' of fold axes to be done: they are discontinuous. The interpreter is not able to complete his visualization in three dimensions, nor is he able to produce a valid contour map, unless the grid size is small enough to give several sections across each positive or negative fold element. All he can produce is a contoured shape which does not conflict with the observed data, but otherwise represents only an hypothesis which can be used to plan the next phase of exploration.

Once the 'joining up' of fold axes and faults is achieved, the contouring process is that of drawing isochrons on flanks of structures which have a fairly uniform direction of slope. The curves should be smooth free-hand curves joining up the points where the contour value cuts the line of traverse. This may be at an observation point, or it may be interpolated by eye or by construction

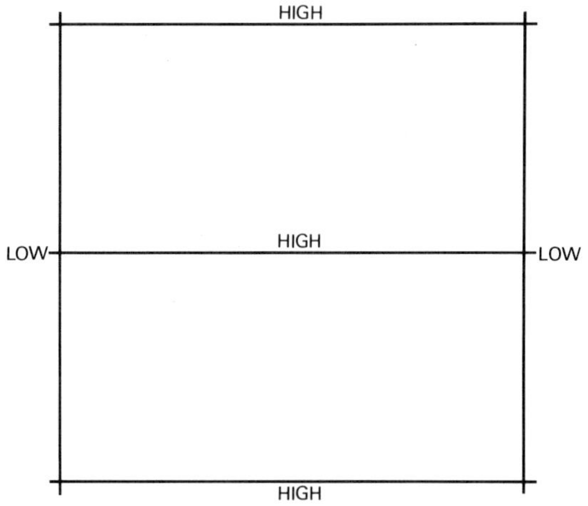

Fig. 28 B. The resolution of the dilemma.
An additional seismic line has been run which leads to the interpretation of a north-south anticline.
Choice of the location of the new line is not critical: but a good guidance is to accept for the moment the most optimistic interpretation, and then to choose a line which will disprove the interpretation if it is wrong, and will give useful information if it is right.
The additional information required is not necessarily new seismic data. Inferences from surface geology or shallow drilling, or interpretation of gravity or magnetic results, may, in the varied circumstances of exploration, establish the structure or lead to the more probable interpretation.

between two observation points. The only other guidance is the shape which the interpreter has in his mind's eye.

Such contour maps in time are valuable first presentations of the results of interpretation. In areas of bold structural relief, the real earth shapes will not be much different, though in flat areas the differences may be profound. In presenting and reporting on contoured time-maps the interpreter should always indicate the sensitivity of the contour maps to velocity changes, together with any velocity evidence available at the time.

There are some other features of contoured time-maps which must always be remembered when using them in exploration. The next most important is that the anticlines are always bigger than in the real earth, and the synclines smaller, as follows from Level 1, chapter 2.

There are some crudities of contoured maps of this class.

The presence of error in the reflection times as they are presented on a seismic section, and the reading errors (which are usually larger), produce small-scale irregularities in mapped reflection time along the line of traverse. These irregularities imply local curvatures in the reflections which are larger than can exist. There is thus a good case for smoothing to reduce the irregularities. This is usually done by the eye of the geophysicist simply by ignoring those data points which would put impossibly sharp bends in his time contours.

The presentation of faults on time-maps is a rather loose approximation. It is clear that in the real earth a normal fault will produce a gap in an horizon: there will be a narrow band on the map where the horizon would not be found by the drill. In the real earth a reversed fault will give a band on the map where the horizon would be found twice. In the seismic earth the reflections always overlap at a fault, either because the fault is reversed, or because it is a normal fault and the reflections are continued as hyperbolas which intersect in cross-section. The compromise is to map the fault by a single line where the reflections or hyperbolas intersect. A consequence is that some of the times mapped are the times on oblique travel paths, not on near-vertical ones. What appear to be structural closures close to normal faults on time-maps are especially suspect, and they must be examined in detail.

The presentation of folds on time-maps breaks down completely in the 'hyperbolas' terrain of para 6.4. All features are convex upwards, and the sharp synclines have the same seismic horizon at three different levels, as in 2.3 If the interpreter can correlate between hyperbolas, he can construct contour maps in time: but the shapes portrayed are so far from real earth shapes that the maps are not of practical value in exploration. The best that he can do is to accept the correlations at the vertices of the hyperbolas, and present the real earth positions of fold axes and faults. Instead of contours he will draw form lines to emphasize the structural elements, and these form lines can be made to accord with the observed times at vertices.

7.4 Isopach maps in time

For a number of exploration problems, the change in thickness of a formation or of some other stratigraphic unit is of importance.

A first assessment of these changes can be made where the unit is defined by seismic horizons at upper and lower interfaces.

The procedure is to determine the difference of reflection times of the boundary reflections at each observation point, to post these values on a map, and contour. The first step is to interpolate faults, which must be in sympathy with those on the boundary horizon maps. The second step is to interpolate lines of maximum and minimum time-difference (or thickness) across the grid squares, and these do not necessarily have any connection with the fold axes, although it will be seen that in some environments there will be a close connection.

In some respects time-isopachs are more crude than time-structure maps.

In an ideal situation, where the bounding reflectors have the same dip, it will be seen that the normals to these two reflectors from an observation point follow the same path through the overlying rocks. The reflection time difference is a true measure of the double travel time through the unit, measured normally – not vertically.

Where there are changes in thickness, the paths in the overlying rocks differ, and the time-difference is no longer an exact measure of the travel time through the unit. The error is immaterial for small changes, but it can be significant for rapid changes where the overlying geology is complex.

The errors become very large where one reflection is carried on a hyperbola, or where the two reflections are recorded from opposite sides of a fault.

Since both upper and lower reflection times involved in the isopach determination are subject to experimental error and reading error, the individual determinations have a larger error. Isopach maps, therefore, can be used only to give broad impressions of thickness change. The fine detail of isopach contoured maps has rarely any geological significance, though larger scale features such as reefs and channel sands can be identified on such isopach maps. Some large errors can be expected locally.

7.5 Building up time-maps of deep horizons

The interpreter will be aware, when he sets out to contour a seismic horizon representing old rocks, that its structure may be complex in detail. The seismic sections show this. Shapes on contour maps are expressions of the interpolation made by the interpreter, controlled only at the traverse lines. There is need to help the interpolation process as much as possible.

It was written in 7.3 that interpolation is most secure when the feature being interpolated is linear or has some other simple form. Extension of this concept leads one to arrange the contouring process so that only those elements with the simplest forms are contoured.

The structure of the shallowest seismic horizon mapped is contoured.This horizon, almost invariably, has the simplest real earth shape, and the simplest seismic expression in isochrons.

Next, the isopach in time between the shallowest reflection and the next deepest mapped horizon is contoured. In the real earth this interval has a variation in thickness which is affected by influences which are relatively simple: original differences in deposition, erosional differences, differences in compaction. The seismic expression adds the effects of simple velocity changes within this interval, and the local irregularities described in 7.4.

Both the first structure and first isopach can be represented by smooth simple contours, although the forms may be quite different.

Determination of the structure of the second horizon down is achieved by adding the two contour maps. At the observation points this simply restores the data values. Within the grid squares, every intersection of a structure contour with an isopach contour gives an interpolated point for the second horizon, which is in accord with both.

By this procedure the more complex structure of the second horizon is built up from the best estimates of two simpler shapes.

Structure of deeper horizons is built up by contouring and 'adding' successive isopachs. The result is a set of structure and isopach maps which are consistent, and geologically plausible.

An alternative approach is to contour independently the structure maps of successive horizons. This involves interpolation across empty grid squares where the geological influences shaping the structure have been complex. It results, sometimes, in structure maps which are not consistent, and isopachs which are not plausible.

7.6 Velocities

Velocities have a two-fold role in seismic work.

In order that the traces of a common depth point gather can be correctly stacked, it is necessary to apply some time corrections to the raw data to counteract the effect of varying source-detector distance within the traces of the gather. These time corrections – normal move-out correction (NMO) or spread correction – are time varying, decreasing with time at a given distance, as is apparent from a simple model.

It has become current practice to express the normal move-out corrections in terms of a horizontally layered earth model, and a restricted source-detector distance, and to determine what the 'velocity' of the seismic waves would be if its behaviour were to approximate the observed step-out times of the reflections. The relationship derived is

$$\Delta T = T_x - T_0$$

$$T_x^2 - T_0^2 = \frac{X^2}{\overline{V}^2}$$

where ΔT is the move-out, defined as the difference between
T_x the reflection time at distance X
T_0 the reflection time at normal incidence
$\overline{V}$ is the time-weighted r. m. s. velocity.

i.e. If there are n horizontal layers above the reflection being considered and for the ith layer the uniform velocity across it is v_i and the reflection time t_i, then

$$\overline{V}_n^2 = \frac{\sum_{i=1}^{n} v_i^2 t_i}{\sum_{i=1}^{n} t_i}$$

where $\overline{V}_n$ is the time-weighted r. m. s. value to the base of the nth layer.

It will be apparent that such a determination of $\overline{V}$ will be affected by all the errors of modelling as well as the errors of measurement. The earth is rarely horizontally layered, and where it is the commercial interest is often not very strong. The gather is rarely restricted to those parts of the spread where the incidence is near normal. The values of $\overline{V}$ do not have any precise significance in terms of the physical properties of rocks.

To those concerned with data processing, this scarcely matters at all, since the interest is not in velocity as such, but in the value of ΔT for the end trace of the spread: it is immaterial if $\overline{V}$ lacks significance, provided that the relationship between T_x and X is maintained with the adopted value of $\overline{V}$ as parameter. There remains the failure of the real earth to provide the hyperbolic relation between time and distance required by the relationship. This, and some other imperfections of seismic modelling, are responsible for the universal failure to reach the calculated improvements of signal-to-noise ratio in stacking. No 48-fold stack shows a 7 to 1 improvement in signal-to-noise ratio.

These imperfections, however, scarcely concern the data processor. With the high multiplicities of data gathering, models and methods far from perfection will yield usable cross-sections.

A quite different order of precision in modelling would be necessary to extract velocity information which has significance as a rock property.

These more precise velocities would be of great value to the interpreter for two uses.

I. If the velocities in the rocks are known, one can calculate the depths corresponding to reflection times. It is only with this depth presentation of the data that the problems of oil and gas migration and accumulation can be considered.

II. The velocity (the velocity of a compressional wave), is an expression of the elastic constants and density of the rock. Velocity may change with changing facies of the formation: for example from sand to shale. It may change with changes in the physical environment of the rock – pressure and temperature. Some of these changes may be reversible, such as the effect of pressure on a clean sand: some will be irreversible, the diagenetic changes in clays and clay minerals such as montmorillonite to illite, or the compaction of chalk. Some changes may be geochemical in origin, such as the cementation of a sand. Finally, some velocity changes are due to changes in fluid content – water, oil, gas, air.

Thus, a sequence of changes in any of these categories can be expressed as a velocity change, other factors being held constant. The reverse process, interpretation of the velocity change, depends on the ability to exclude some factors from consideration on grounds of geological plausibility.

This preamble is enough to indicate the great interest which precise velocities hold for the interpreter. It will suggest, also, that much of the velocity data available to-day is not sufficiently precise for the use of the interpreter. However, the best data, especially if a suitable smoothing function is applied, will already yield a meaningful interpretation of the major velocity features.

7.7 Conversion to depth and thickness

Analysis of step-out times yields values of $\overline{V}$, as previously described, for various levels. What is required for depth conversion is V_a, the average velocity; and v_i, the interval velocity, is required for the thickness conversion.

It will be clear that the plot of $v_i^2\, t_i$ against T_0 is a straight line – since v_i is taken to be constant for the interval. Thus the plot of $\overline{V}^2\, T_0$ against T_0 is a series of straight line segments, the slope of each segment affording a measure of the corresponding interval velocity. For a given value of X – usually corresponding with the most distant detector a plot is made of

$$\frac{X^2 \cdot T_0}{T_x^2 - T_0^2} = \overline{V}^2\, T_0 \text{ against } T_0$$

The data available will be corresponding pairs of T_0 and T_x values on identified reflections within the gather: and this will provide isolated points on the graph of $\overline{V}^2\, T_0$ against T_0. Joining these by straight line segments is the assumption of constant interval velocity: and for the interpreter this assumption is broadly justified if he can find no prominent reflection within that interval.

The differences of successive T_0 values give the series of t_i values, and the slopes of the graph segments give the values of v_i^2. With the values of t_i and v_i one can compute the value of

$$V_a = \frac{\sum_{i=1}^{n} v_i\, t_i}{\sum_{i=1}^{n} t_i} \quad \text{for the average velocity to the } n^{th} \text{ reflector.}$$

This average velocity converts reflection time to depth, or more strictly normal distance.

The individual values of v_i are the interval velocities needed for thickness conversion.

Successive values for V_a and v_i for a particular horizon along the line of traverse, derived by this means, will generally show a considerable scatter: and it is important that the values be smoothed before any conversions are attempted.

Discussion so far has ignored the need to correct depths from normal incidence distances to true depths; and to convert (if need be) from normal thicknesses to vertical thicknesses. These subjects are dealt with in a later discussion on migration.

7.8 Use of well data and other geophysical data

A wide variety of data is yielded by a well: and all of it, eventually, must fall into place in the complex geological picture to which the seismic survey has contributed.

Sample study, aided by palaeontological and mineral determinations, will establish the stratigraphic units in the well from cuttings and cores. Correlations based on electric logs will extend the identifications to neighbouring wells.

Velocity logs and calibrations in the well will enable calculations to be made of the theoretical reflection time of geological horizons whose depths have been determined in the well. Comparison with a seismic section leads to identification of a reflection as corresponding with that geological boundary.

Usually, the time on the seismic section will be larger than that calculated, for several reasons. The time measured in the well is from the initiation of the source pulse – the time break – to the leading edge of the arrival at the well geophone – the first break. Reflection time on the cross-section is measured to a peak or trough which can be used, which is some time after the leading edge: the leading edge of a reflection is usually obscured by noise or by other seismic arrivals. Analogue filters used in recording the reflection also add delay to the measured reflection time.

Magnetic and gravity surveys were often run as a preliminary to seismic survey in earlier times. Now, especially in marine surveys, they can be run concurrently. The principal use of this extra data is to help solve special problems. For example, a mobile formation which has formed a dome or a piercement structure might be a salt or a shale. The difference in density is large enough for the gravity results to discriminate between them. Such a feature could equally be an intrusive igneous rock. Its density will be still higher than that of the shale; and a basic intrusive will be denser and more magnetic than an acid one.

These are examples where the gravity or magnetic contribution to the interpretation is clear and precise. More generally, the seismic, gravity and magnetic observations must be reconciled. If suitable values of density and magnetic susceptibility can be assigned to the geological units outlined by the reflections on the seismic section, then it is possible to calculate the gravity and magnetic fields generated by this model, and to attempt the reconciliation.

7.9 Reporting the results

The purpose of seismic interpretation is, in the vast majority of cases, to further exploration for oil. Decisions will be taken by exploration managers. It is an important function of the seismic interpreter to communicate his results: by written reports, by discussion and by demonstration.

The interpreter may work on a given task for a few days, or for many years. His work requires that he handles, understands, and pieces together a vast amount of detail. Yet his reporting must be selective: it is not of value to reproduce all the detail. To be selective, he must understand the exploration problems of the area that he is working, otherwise he cannot exercise judgement in selecting and summarizing his results. The interpreter is, as it were, the bridge between the world of physics, mathematics and computing which is concerned with seismic data gathering and processing, and the natural world of geology, geochemistry, fluid migration and the closely allied world of commercial assessment and risk taking in the petroleum industry.

Of prime importance in an interpretation report is the summary – or 'executive summary' in a commercial report. This, in about a page, should summarize what has been done, and outline those findings of the interpreter, which, he believes, will deserve consideration in the next phase of exploration. This summary would include structures or stratigraphic configurations which might trap hydrocarbons; and any direct observations bearing on accumulation – structurally high reflections in geologically plausible locations for accumulation, for example. Of great importance in this summary is the measure of confidence which the interpreter has in his findings. It must be clear to the reader whether the interpreter considers the results given in the summary are established by the survey, so that decision-taking can be based on them. If they are not established, then recommendations for further work must be made, on the results of which the interpreter could report firm conclusions.

The interpreter should report both classes of findings – the established features and those needing more work. Occasionally the latter turn out to be of more value than the original objective. Examples of these poorly established features are the observations of dip contrary to regional dip on the edge of the survey, the ill-defined possible reef, and so on.

The body of the report should be much fuller: and it should be written with the needs of the user in mind. For example, the executive must be able to find detail on the points of interest in the summary. A geophysicist must find what he needs to plan an extension to the survey, or a more detailed local survey within the area. The geologist may approach the report with nothing but surface geological observations, and outcrop information from nearby hills: or he may have a wealth of subsurface information. It is within the framework of the seismic report that he must envisage the geochemistry of hydrocarbon formation, the mechanics of its migration, and the features of reservoir and cap rocks which control its emplacement at the present day.

Other references may be made also. In a wild-cat area the seismic interpretation report may offer the only information on which to plan a casing programme for a first well. A marine seismic survey may offer the first assessment of the nature of the sea-bottom on which a rig is to sit – although in many cases special investigation of the site is made by direct methods.

What is not required in the report is routine descriptions of cross-sections and maps. In general, the documents communicate their message better than the written words describing them.

LEVEL 2

8 The seismic pulse

8.1 An examination of pulse form

There is evidence that in the transmission of sound through a homogeneous medium, high frequencies are attenuated more strongly than low frequencies. In the Pierre Shale, where ideal conditions are approached, it has been shown that the returning reflection pulse from a small explosion has a simple shape which broadens with travel time (McDONAL et al. 1958).

Lengthening the path, therefore, lowers the central frequency; and if a spike is input the observed reflection will be a pulse of finite duration – an 'absorption pulse'. In the real situation where the input is itself a pulse the reflection will be this input pulse convolved with the absorption pulse.

This effect can be a changing degree of resolution. A low frequency pass-band may slur two closely spaced reflections into a single trough, while a high frequency pass-band resolves this into two troughs which can be followed separately. An effect which looks similar to the interpreter's eye arises from the use of narrow pass-band filters, markedly less than one octave. This lengthens the pulse form and the spreading may increase the number of troughs in the pulse form. The results described do not arise as a result of the filter choice in processing: the cross-section presents the cascaded filter effects of the earth, recording and processing systems, and the splitting or combination of troughs is determined by the resulting equivalent filter. It is the splitting of the reflection trough from these causes that obscures and is confused with the 'splits' of geological origin and their interpretation is made more difficult in consequence. (See Level 1 para 1.8).

Thus the correlated Vibroseis pulse is a Klauder wavelet (the autocorrelation of the sweep), convolved with an 'absorption' pulse, and the returning reflection pulse in a marine survey is the source signature convolved with an 'absorption' pulse. All these are true under the same idealized conditions.

These observations are helpful in understanding a part of the seismic process, but they have little practical importance to the interpreter. The reason is, that, as we have seen in chapter 3 Level 1, there are many later influences on pulse form – some convolution type processes, some additive processes – and some irregular and illogical processes such as normal moveout correction. These affect the pulse form so much that the idealized pulse form is a rarity.

Under standard conditions the form of a pulse may have the imprint of the microstratigraphy around the reflector so that the pulse form contributes to seismic correlation. Surface changes, however, may be so great that they over-ride the stratigraphic influence.

The greatest power for changing pulse form resides in the tools of data processing. For example, choice of band-pass filter can split a pulse into several pulses, or merge serveral into one. Pulse-shaping filters can make them triangular or lop-sided or almost any desired shape. They can even reduce amplitudes so that the reflection drops below noise level and is effectively lost.

In the sections that follow the effect of pulse form change on some interpretation procedures will be studied, as will a few kinds of change which are themselves interpretable.

8.2 Pulse form in continuity

The concept in the use of seismic continuity is that there is a direct correspondence between a seismic pulse, and a short section of the microstratigraphy across a lithostratigraphic horizon. In practice this is usually simplified to a correspondence between some element of the pulse, usually peak or trough, and the lithostratigraphic boundary itself.

In this simplification one sees a source of difficulty. Changing stratigraphy at the boundary will change the pulse form, and the chosen pulse element, peak or trough, may move relative to the stratigraphic element selected as the boundary.

The changes are usually of a minor sort, and are generally overshadowed by changes in the time of peak or trough resulting from the impact of noise, or from the recording or processing parameters. Field conditions may force a change of technique in recording. In processing, the phase shifts introduced by band-pass filters, or by pulse-shaping filters such as 'spiking deconvolution', are well known. Less familiar is the shift resulting from incorrect velocities on stacking. This can reach a quarter of the period of the reflection.

It is seen that there are some in-built sources of error in the use of seismic continuity. Those in the recording and processing parts of the system are made smaller as the improving technology reduces noise and processing distortion.

Some tests have shown that the standard error of timing good reflection troughs on repeat observations under different conditions lies between 1 and 1½ ms. A standard error of 1 ms in individual reflection times is an almost attainable ideal. Much current interpretation discards this essential precision of the seismic method, and reads reflection times to the nearest 5 or 10 ms, thus introducing the largest error contribution as this observation error.

By using high frequency reflections, and suitable smoothing prodecures to the accurately read reflection times, it is certain that the best reflections can be timed to an error of less than 1 ms. These precision procedures are aided by using a digitizing table to digitize the interpreter's marked horizon on the section. This data will have errors of the order of 5 to 10 ms, but it can be used to re-enter the plot tapes which produced the cross-sections, and to search for the peak or trough, either to the nearest sample, or by interpolation procedures to higher accuracy.

Perhaps the greatest difficulties with changing pulse form arise when the cross-sections of the area of the interpretation are recorded and processed by different organizations, or at such widely spaced times that techniques are different. At the junctions there is often no straight comparison of pulse forms. Neither is it valid to select the nearest trough, and attempt to set up a time-difference which will make the reflection times of one technique equivalent to those of another. The geological sequence changes from place to place, and so the time-shift between records taken with two different seismic systems cannot be expected to remain constant. See also Level 1 Section 1.7.

8.3 Pulse form in correlation

All the features which bear on the changing pulse form in continuity affect the pulse form in correlation. The difficulties are increased by the absence of intermediate forms which enable one to link pulses of different form by continuity.

The changing pulse forms in continuity lead to small reflection time errors as the timed element changes its position slightly with respect to the detailed stratigraphy. In correlation the changes in pulse forms can result in a 'blunder' – correlating one pulse with another which is an expression of a different part of the geological sequence.

One must add that correlation is usually established on a reflection sequence, not on an individual reflection. There are cases, usually in the lower parts of sections, below an unconformity, where only few reflections are to be seen at intervals along the line. Correlation must then rely on pulse form.

8.4 The pulse form at unconformity

The pulse form can be regarded as the convolution of a 'unit' seismic pulse with a part of the log of reflection coefficients. Typically, the part of this log involved shows a principal spike, usually corre-

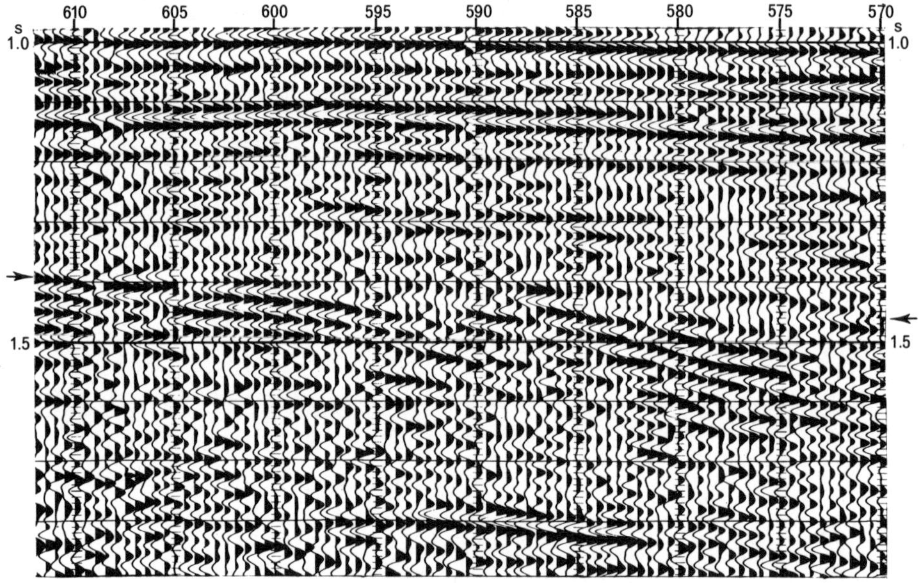

Fig. 29. An unconformity, marked by arrows. It is recognized by the angular discordance between upper and lower formations, seen most clearly in the view along the section from left or right, with the eye close to the plane of the paper. There is no continuous reflection, in the usual definition, at the unconformable surface.
Instead, there is a reflection which sometimes changes phase abruptly, as between 604 and 605. In other places the amplitude of the reflection drops to a low level, and it is lost in the noise. This can be seen between 587 and 597.
Such changes are due, almost certainly, to the changing lithology, and related velocity and density changes, of the successive members of the lower formation that are overstepped by the basal member of the upper formation. The response at the unconformity is of value in effecting seismic correlation within the lower formation.

sponding to the significant lithological and velocity change which defines the boundary. At the leading end there are minor features of the log expressing the stratigraphy of the lowest part of the upper formation. At the trailing end are similar features from the topmost part of the lower formation.

It will be seen that at an unconformable boundary showing overstep, the features from the lower formation will be changing along the line of traverse. As a result, the pulse form will change – particularly its trailing end. Other relationships at unconformity will produce other kinds of change.

The result is a 'reflection' of highly variable character. It shows an alignment, but rapidly varying shape. This feature can sometimes be used to establish the position of an unconformity. It is also a feature of the reflection which can be used in making correlations. Fig. 29.

There are cases where a reflection at an unconformity appears to be unaffected by the changing lithology of the underlying formation. Such, for example, is the 'Base Tertiary' reflection in the southern part of the North Sea. It is a fair inference that such a reflection is not generated at the interface between Tertiary and Cretaceous, but by a basal Tertiary complex. Changes in such a reflection may occur over a wide area of search: but they will be due to facies changes within the basal complex, not to the effects of change in the lower formation.

9 Geometry: Migration

9.1 Three-dimensional migration

Hagedoorn (1954) has shown that with a simple model, a simple migration procedure in three dimensions can be established.

The model is that of a reflector overlain by a uniform rock – i. e. one of uniform velocity. The process is to convert the normal reflection times to one-way path lengths by multiplying them by V/2, where V is the uniform velocity referred to. These path lengths are plotted at the surface point at which they were observed, and the values contoured; i. e. the path lengths are treated as depths would be. A line of vertical section at right angles to a contour contains the observation point, the true reflecting point, and the true normal to the reflecting surface at the reflecting point. It becomes a simple matter to migrate the contour point to its new plan position and new depth. Alternatively, by making the assumption that the dip is locally constant, the migrated position of the contour can be found.

These relationships are shown in figs. 30 A and 30 B.

As a numerical example, take

$$x = 2500 \text{ m}$$
$$d_1 = 1000 \text{ m}$$
$$d_2 = 2000 \text{ m}$$

Then

$$\alpha = \tan^{-1}\frac{2}{5} = 21°\ 48'$$

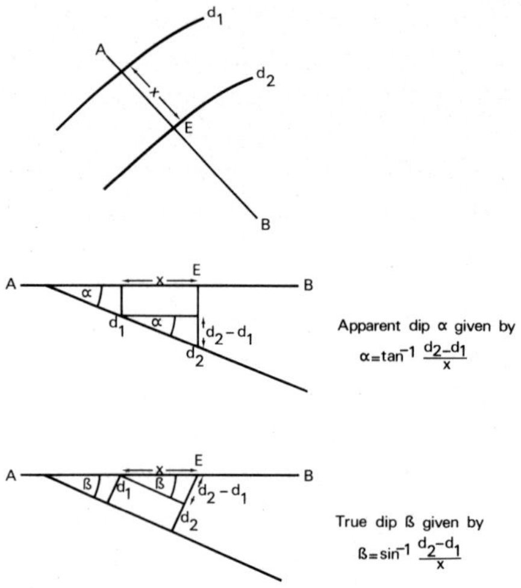

Fig. 30 A. Migration relationships of contours.

$$\beta = \sin^{-1}\frac{2}{5} = 23^\circ\ 35'$$

$$\begin{aligned} CD &= 2000 \cot 23^\circ\ 35' \\ &= 2000 \times 2.2907 \\ &= 4581 \end{aligned}$$

i. e. a displacement of 419 m up dip, since

$$CE = 5000.$$

This represents the displacement of the 2000 m contour and

$$\begin{aligned} b &= 2000 \sin 23^\circ\ 35' \\ &= 2000 \times 0.4 = 800 \text{ m.} \\ c &= 2000 \cos 23^\circ\ 35' \\ &= 2000 \times 0.9177 \\ &= 1835.4 \text{ m.} \end{aligned}$$

I. e. the reflection point observed from E, which falls on the unmigrated 2000 m contour, migrates with a displacement of 800 m up dip, and a decrease in depth of 165 m.

Where the material above the reflector is not homogeneous – the typical case in practice – further refinement is neccessary.

It can be shown that, to a first approximation, the required migration can be calculated using V_a, the average velocity to determine vertical displacements, and $\overline{V}$, the r. m. s. or stacking velocity to determine lateral displacements.

However, these methods become less and less accurate as the geological complexity increases, and one is driven eventually to modelling, as discussed in a later section.

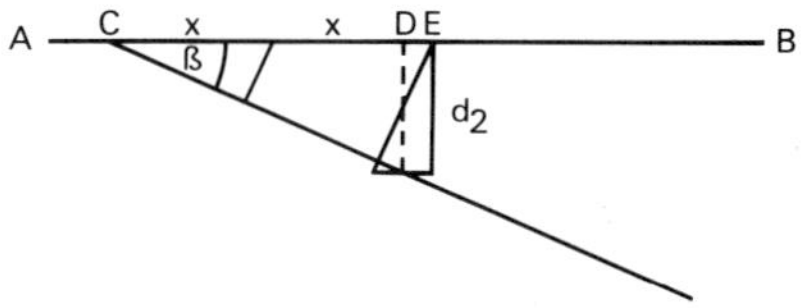

Displacement of the contour is given by CD where $CD = d_2 \cot ß$

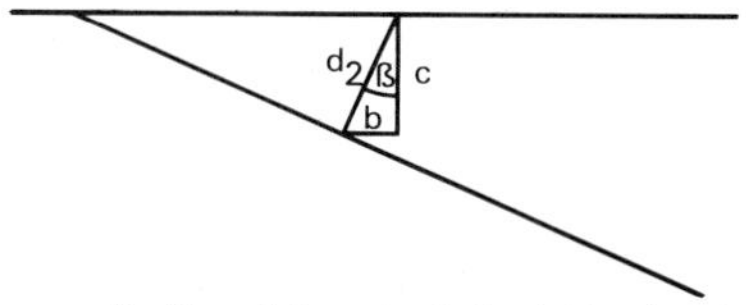

Position of the migrated point is given by
$b = d_2 \sin ß$
$c = d_2 \cos ß$

Fig. 30 B. Further migration relationships of contours.

9.2 Two-dimensional migration

The three-dimensional migration procedure of the last section is applicable only to a seismic horizon which has been traced, timed, converted from reflection times to path length, and contoured. The resulting output is a contoured horizon in its true earth position.

All of the fine detail seen on cross-sections between these contoured horizons is lost, in the sense that it does not appear on the migrated, three-dimensional presentation. For many purposes it is important to examine this detail in relation to the principal reflections in its true or migrated position with respect to the cross-section.

These considerations lead to the migrated cross-section. First will be discussed the migration of a seismic section known to be at right angles to the strike; and then the migration of a section oblique to the strike.

The problem for the section at right angles to strike is illustrated in fig. 31.

Fig. 31 a represents the real earth. The problem set is to obtain some data on the reflector at the point P. The search for this data will be carried out by making observations at points O_1 O_2 . . . until a normal incidence observation is found, i. e. until O_rP is at right angles to the reflector at P.

This same search routine is illustrated on a seismic section in fig. 31 b. The possible positions where the data for point P may be found lie on a hyperbola. The geometry of fig. 31 is that of a common depth point gather. What corresponds with the source-detector distance is 2X, where X is the distance from the search point O (vertically above P), to the point O_r at which normal incidence to the reflector at P is achieved.

The relationship on which most currently available migration programmes are based is thus

$$T_s^2 - T_m^2 = \frac{(2\,X)^2}{\bar{V}^2}$$

where T_s is the recorded seismic arrival time from point P
T_m is the corresponding time on the migrated section to point P
$\overline{V}$ is the appropriate r. m. s. velocity
X is the offset distance of O_r as described above.

There are a number of ways of conducting the search and presenting the results: and some alternative theoretical approaches also, though the concern here is essentially to discuss the impact on the interpreter.

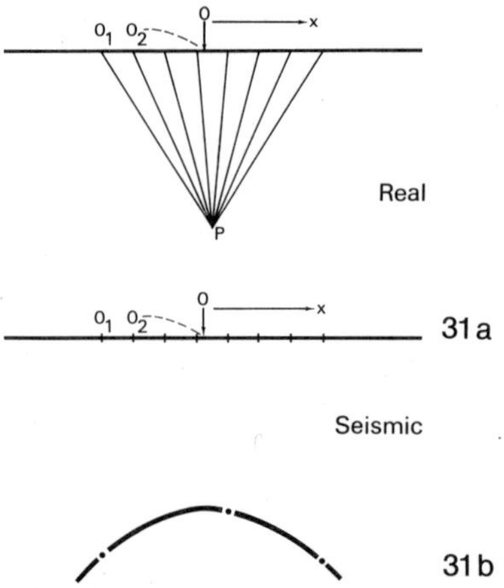

Fig. 31. The search for the seismic data pertaining to a real earth point P (which is the seismic data one wishes to place at the migrated section point P corresponding to it).
If the reflector is horizontal, the required data will be observed from the surface point O, vertically above P. If the reflector is not horizontal, the required data will be observed at one of the points O_1, O_2 ... O_r ... The actual observation point will be O_r if O_rP is normal to the reflector.
This same search on the seismic section is seen to be for a data point lying somewhere on the hyperbola.
Note that this diagram is very similar to that which would be drawn to illustrate a common depth point gather. It is not quite identical, as will be seen later.

In areas where the structural features are parallel, the dips are not too extreme, and the section is at right angles to strike, the migration process works as forecast. Anticlines and synclines are restored to their correct proportions (though still with the vertical axis in time). Synclines which show reverse curvature on the seismic section because their centres of curvature have fallen below the surface have their true curvature restored. The fundamentals of the change from 'seismic' to 'real', as used in the figures of chapter 2, Level 1, have been accomplished.

In addition, hyperbolic features arising from the truncation of reflectors by faulting or in other ways are shrunk to a point by the migration process.

The value of the process for the interpreter lies in part in this restoration of folds to a more naturalistic presentation. The more valuable contribution is in the interpretation of fault structures. Anticlinal traps are not moved in position by migration – only narrowed; but fault traps in dipping formations are moved up dip by the migration process, and the closer determination of position is valuable.

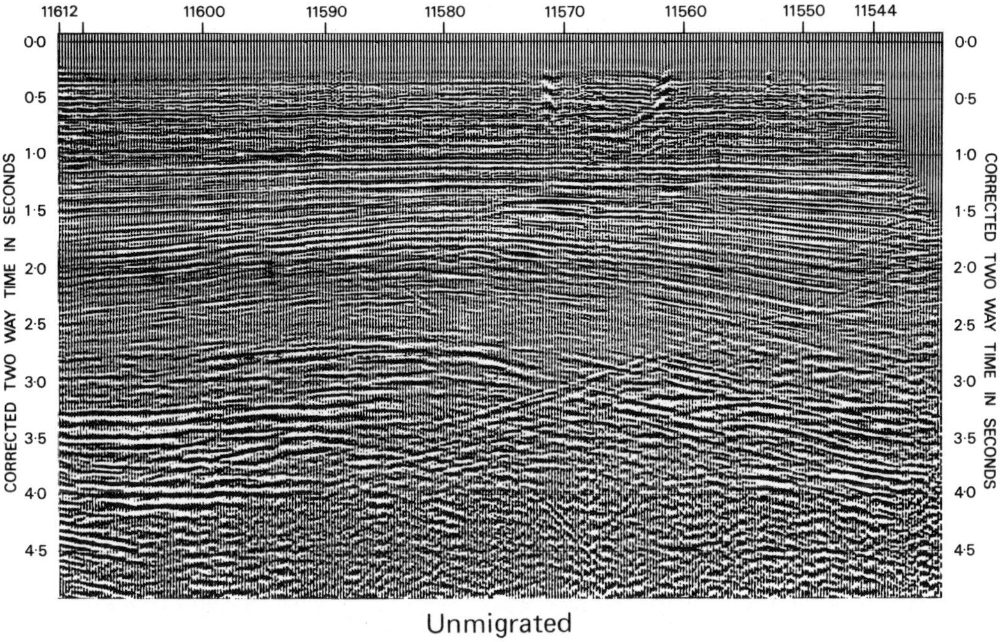

Fig. 32 A. A seismic section in an area with considerable faulting. The interpreter is left with the impression of faulting, but it lacks precision: and most interpreters would hesitate to draw firm fault lines.

A further benefit in migrating a faulted area is seen particularly in regions of impersistent stratigraphic units and reflectors, as is typical of a delta. The impersistent reflections are obscured by weak hyperbolas arising at their truncations, and neither reflections nor faults can be traced with confidence. The effect of migration here is to remove the hyperbolic interference, and so improve the clarity of the reflections. At the same time the fault trace is revealed as a clear line on the seismic section at which the reflections appear to be 'butt-jointed' either with a small displacement, or with a somewhat different reflector sequence on the two sides in the case of a growth fault. These features are well illustrated in figs. 32 A, 32 B and 32 C.

Where the structural features are parallel, and the line of traverse is not normal, but somewhat oblique, a fair migration can be effected if the velocities are 'adjusted'.

Fig. 23 illustrates that if the line of traverse is at an angle α away from the ideal normal-to-strike direction then the hyperbola from a fault truncation or any other linear buried focus is stretched horizontally.

The T_x value which is found at a distance X in fig. 23 a if found at $X' = X \sec \alpha$ in fig. 23 b Level 1.

One can write

$$T_x^2 - T_0^2 = \frac{X^2}{\bar{V}^2} = \frac{X'^2 \cos^2 \alpha}{\bar{V}^2}$$

or

$$T_x^2 - T_0^2 = \frac{X'^2}{\bar{V}^2 \sec^2 \alpha}$$

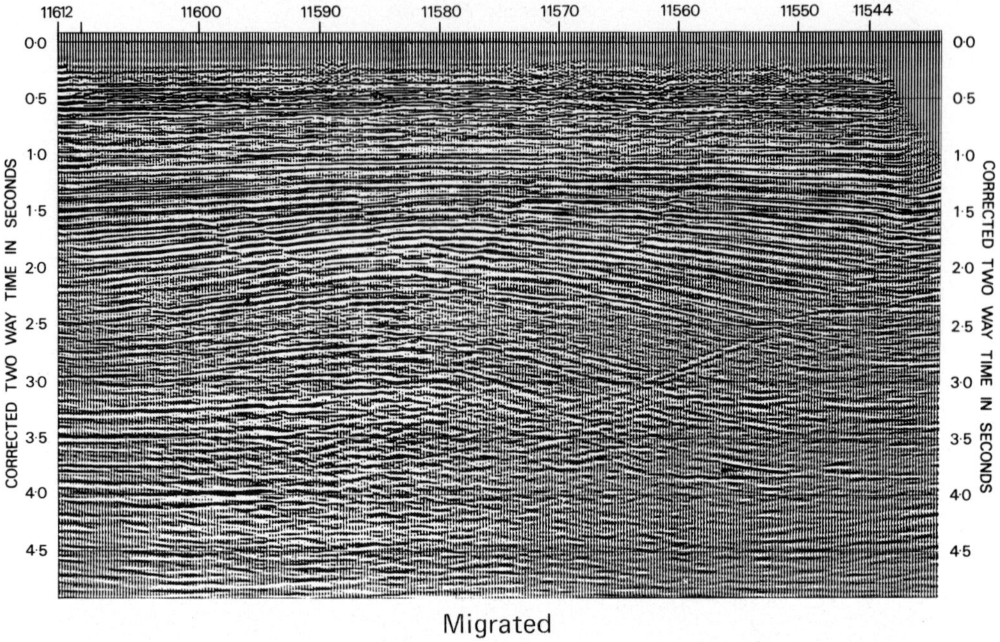

Fig. 32 B. The same data as fig. 32 A migrated. It shows an important growth fault at the extreme right, very shallow, hading to the left with a decreasing hade, and passing into a fault plane reflection with only a gentle hade.

Above the growth fault the beds have been folded into an anticline which is cut by a number of faults, all normal faults, which hade towards the anticlinal axis, and step down towards it. These faults all show as clear lines across which is a small displacement, so that the appearance is that of wood which has been butt-jointed. There are four faults on the left, three on the right.

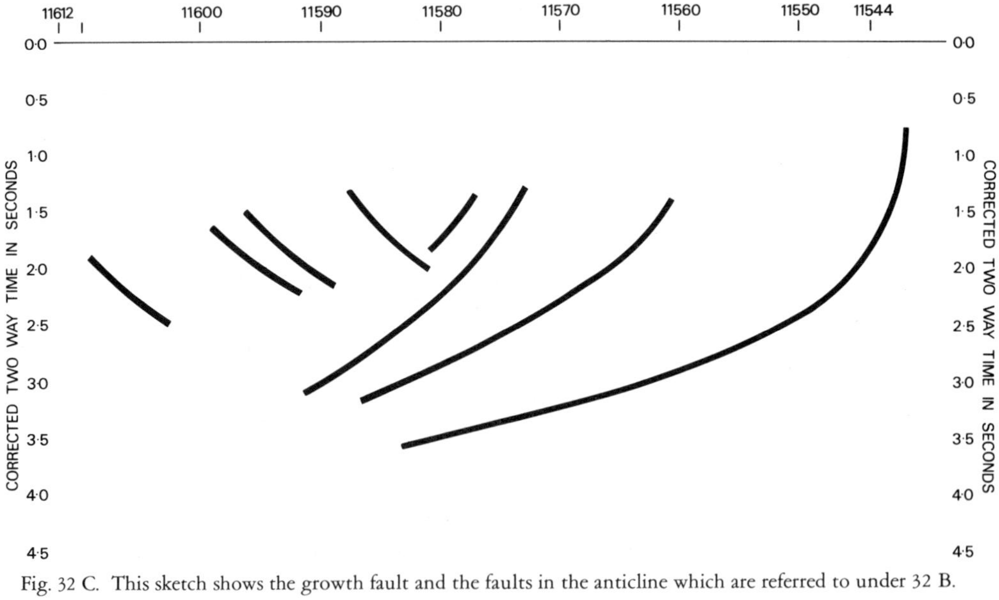

Fig. 32 C. This sketch shows the growth fault and the faults in the anticline which are referred to under 32 B.

So that all the manipulations can be carried out using a sequence of velocities $\overline{V}$ sec α in place of the observed values.

It will be apparent that where features along a line of traverse have different strike directions, no simple migration process is possible. Such cases require modelling, as described in section 9.9.

Other sources of difficulty arise where the seismic sections contain alignments which are not primary reflections, i. e. multiples, refracted reflections and reflected refractions: the last two are especially evident in strongly folded areas. These other alignments are merely distorted by the migration process, and tend to mislead the interpreter. Equally disturbing is the existence of shadow regions in complex structures, from which no reflections can be observed at surface – though occasionally these shadow regions can be probed by reflected refractions or wide-angle reflections.

Where these geological complexities obscure the results of migration, the interpreter must rely on the qualitative interpretation which results from application of the concepts of chapter 2, Level 1; or he must resort to modelling, and by trial and error arrive at an acceptable solution.

9.3 Velocities involved in migration

From the analogy drawn between the geometry of migration and the geometry of the common depth point gather in fig. 31 it can be inferred that the stacking velocity is also the migration velocity. This is broadly true, but it is subject to some practical restrictions. The modification to deal with oblique lines has already been discussed in the previous section.

The identity is rigidly true only for horizontal bedding, which is of no practical interest. As the dip increases, a difference develops between the diagram for the CDP gather, and that for the migration search.

At wide source-detector separation, the reflecting point in the 'CDP' gather moves up dip. This results in smaller reflecting times for the outside traces than if the path were to the normal incidence reflection point: in consequence the stacking velocity is higher, and is no longer equal to the r. m. s. velocity or $\overline{V}$. On the other hand, the point P for the migration search, (see fig. 31 a), must remain fixed, and the corresponding reflection times are larger, the velocities smaller. In an appendix to this section are given the geometrical relationships, which can be used to assess the magnitude of the departure in specific cases. This is from P. E. F. Hawkins (private communication).

It can be demonstrated that in a complex geological situation, with steep dips involved, the 'velocity' to be used in the formula of section 9.2 varies with the dip significantly, so that the simple migration concept of that section no longer works. Moreover, the value of the 'velocity' so derived may depart very far from realistic values of rock velocities, and the whole process becomes unreal.

The papers cited by Gardner, French & Matzuk (1973), French (1974), have valuable discussions of the relation of velocities to the migration process, and point the way to further developments.

9.4 The effect of obliquity of line of traverse to the full dip line

One effect has already been discussed in section 9.2. Hyperbolic features, generated by any buried focus, are stretched horizontally on cross-section as the departure of line of traverse from normal to strike increases.

A more marked and more confusing effect arises in an area of complex pitching structures. The result is then to superpose on the seismic sections arrivals from several different directions, and the result may be a complete confusion. These conditions may defeat the interpreter if he attempts to interpret an isolated section. Each system of reflections may show sections of continuity, with areas of confusion between them. Without further evidence one cannot tell which continuous reflections are related to which others, nor can one tell the directions from which they arrive. Sometimes, the examination of other sections, with better orientation, will reveal enough of the structure to enable one to interpret these complex sections with arrivals from various directions – known familiarly as 'side-swipe'.

9.5 Some considerations for complex areas

Two aspects of complex structure have been touched on in previous sections.

The first is that the approximate equality of stacking velocity, r. m. s. velocity, and migration velocity, and the relationship of this velocity to the true velocities in the rocks of the sequence, no longer holds. The values of these velocities begin to diverge, and the values become data-dependent; the 'velocities', in other words, depend on dip and other structural parameters.

The second aspect already mentioned is the confusion which appears on some seismic sections, where the normals to the reflectors from the line of traverse can probe different structural elements, and the seismic arrivals are superposed on the seismic section.

A further aspect introduced here is that of shadow-zones which appear in complex areas. No reflections are recorded from some parts of a reflecting horizon: and so no evidence exists as to the nature of these parts of the structure. The complexity which produces these 'shadow areas' may result from complex structure or complex sedimentary conditions. Associated with 'shadow areas' are usually curious arrivals which can be identified as refracted reflections; or as reflected refractions.

It can be shown that with certain types of velocity distribution a reflection from any interface can be recorded at the surface. This is true even for beds which are overturned, though for most models it excludes the limiting case of a reflection from the underside of a horizontal bed – whether overturned or in its correct position. This holds true, for example, for a model in which the velocity increases linearly with depth, and the isovelocity surfaces are horizontal. It may be approached, in practice, in cases where the beds overlying the reflector are of rather uniform lithology, and the velocity is determined largely by the loading, so that it increases smoothly with depth. Geologically, it would be unusual to find a steep or overturned reflecting horizon in this environment. Geophysically, reflection times from steep or overturned reflectors could be very large with this model. They have rarely been recognized or identified on seismic section.

What happens more frequently is that there is a velocity inversion somewhere in the overlying sequence; a bed has a higher velocity than the one beneath it. This bed may form part of a nearly horizontal sequence overlying the folded rocks, or it may be involved to a greater or lesser extent in the folding.

Fig. 33 illustrates the horizontally bedded sequence diagrammatically. It will be observed that the high velocity bed functions as a leaky wave-guide. The seismic section shows the curve, almost a hyperbola, which represents the buried focus. At the point where the wave-guide takes over, the 'hyperbola' is continued as a straight line tangential at that point. It can easily be demonstrated that

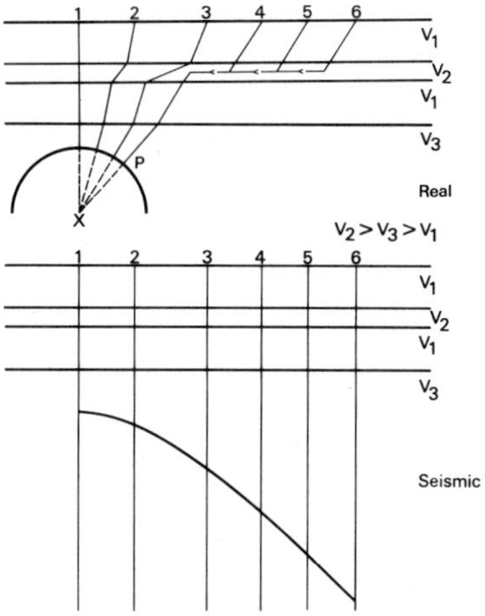

Fig. 33. Illustrates that normal incidence rays controlled by buried focus (in this case the centre of curvature of an anticline), can be caught in a higher velocity bed, which then functions as a leaky wave-guide.
The effect on the cross-section is that the hyperbola passes into a straight line where the wave-guide takes over. More correctly, the curve is not a hyperbola because of refraction and the line is not straight, but portrays the structure of the V_2 refractor.
At first glance the seismic section appearance is very similar to that of Level 1 Fig. 19, so the essential differences are tabulated.

	Reflected refraction	Wave-guide refraction
(I)	Hyperbola changes phase	Hyperbola does not change phase
(II)	Centre of curvature at vertex of hyperbola	Centre of curvature below vertex of hyperbola
(III)	A reflection tangential to hyperbola at point of phase change	No such reflection

It is to be noted that there is no seismic data for the steep limb of the anticline, below the point of normal incidence for the ray in the wave-guide point P. This limb cannot be explored by the seismic method: the bed of velocity V_2 acts as a complete screen.

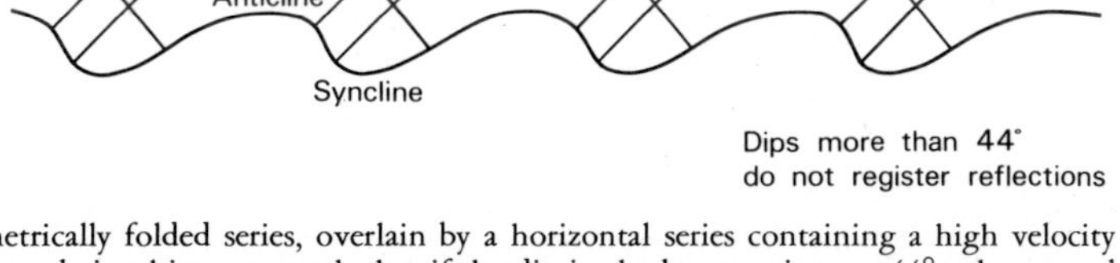

Fig. 34. A sketch of an asymmetrically folded series, overlain by a horizontal series containing a high velocity bed. In this example the velocity relationships were such that if the dip in the lower series was 44°, the normal incidence wave would be channeled in the wave-guide. The folds were such that the gentle limb of the fold gave a normal seismic response; most of the syncline gave a convex upward response since the centres of curvature were subsurface; and the steep limb exceeded 44°, so that the conditions of fig. 33 were reproduced twice on each steep limb.

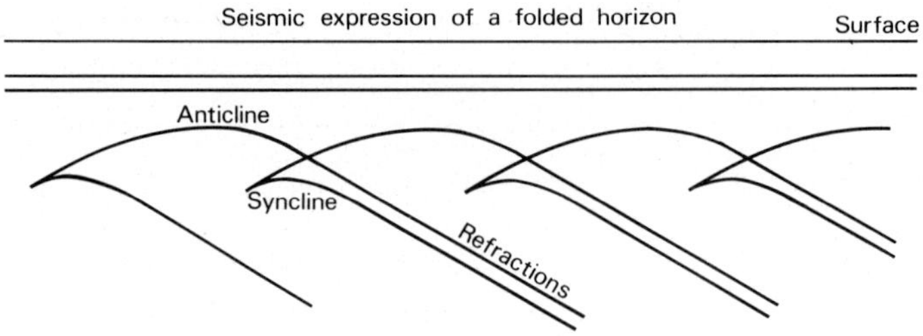

Fig. 35. The seismic response of fig. 34. Note the broad, gentle anticlinal response and the narrow, more strongly curved synclinal response. Each steep limb puts two normal incidence rays into the wave-guide, and in a sense the gap between the two refractions from the limb represents the unexplored part of the limb.
The example of figs. 34 and 35 is taken from material with detailed velocity control.

Fig. 36 see folder.

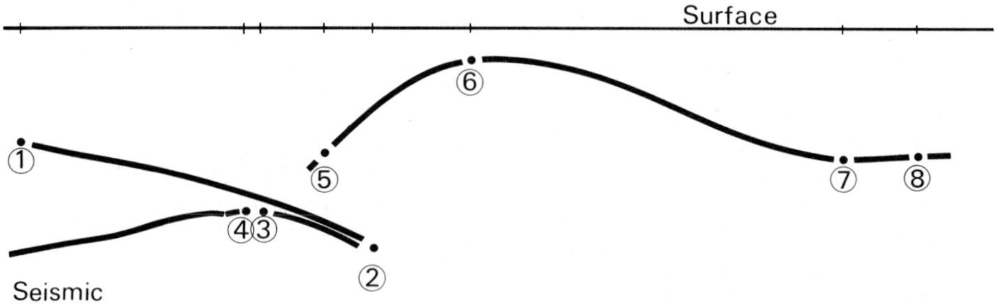

Fig. 37. This reproduces schematically the essential alignments for a single reflector which are recognised on fig. 36.

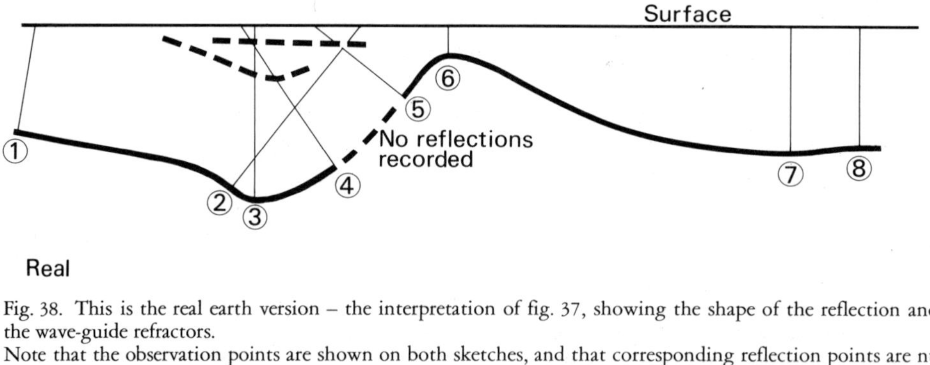

Fig. 38. This is the real earth version – the interpretation of fig. 37, showing the shape of the reflection and of the wave-guide refractors.
Note that the observation points are shown on both sketches, and that corresponding reflection points are numbered. No seismic observation can be made between 4 and 5. Further consideration suggests that this steep limb is probably a good deal steeper than this sketch shows, and that it may even be overturned.

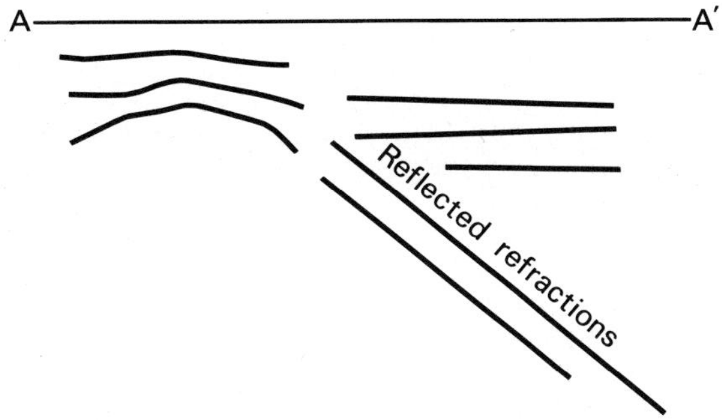

Fig. 39. Sketch showing the elements to be expected on seismic section across the margin of a great reef. The reef is on the left, with off-reef deposits, almost horizontal, on the right.
Refractions along bedding planes in the off-reef deposits, with appropriate velocity contrasts, are reflected back from the steep reef face. These reflected refraction arrivals contain information only about the off-reef refractor.

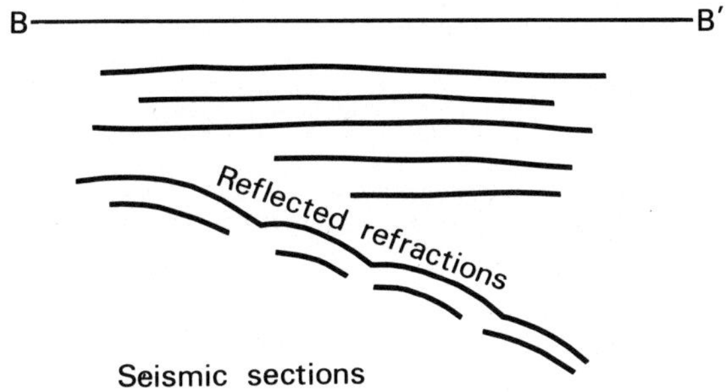

Fig. 40. Sketch of the corresponding section nearly parallel with the reef face.
The general course of the reflected refractions contains information about the strike of the reef front. The hyperbolic features that develop along it contain information about the centres of curvature of the reef front as seen in a horizontal section. The reef fronts are typically marked by large-scale vertical corrugations.
If the off-reef sediments were quite horizontal, it can be seen that the reflected refractions from several different off-reef refractors would approximate to contours of the reef face: and after migration (in a horizontal plane) they would approximate it still more closely.
Stated more generally, the cross-section of fig. 39 gives information on the off-reef structure and a few points on the reef-face. The cross-section of fig. 40 gives much more information on the reef face, but needs that of fig. 39 to help interpret it. Provided that enough off-reef refractors can be found, a seismic method of exploring the reef face can be devised.

if the high velocity formation is much thicker, a recognizable curvature may remain on this continuation.

It will be appreciated that the hyperbolic part of this seismic arrival contains information about the folded reflector below the unconformity. There is no information deeper than the point of normal incidence of the ray involved in the wave-guide: below that point is a shadow zone. The straight line continuation of the hyperbola contains information only on the shallow refractor which serves as wave-guide. If the shallow refractor is folded the 'straight line' takes on the fold shape.

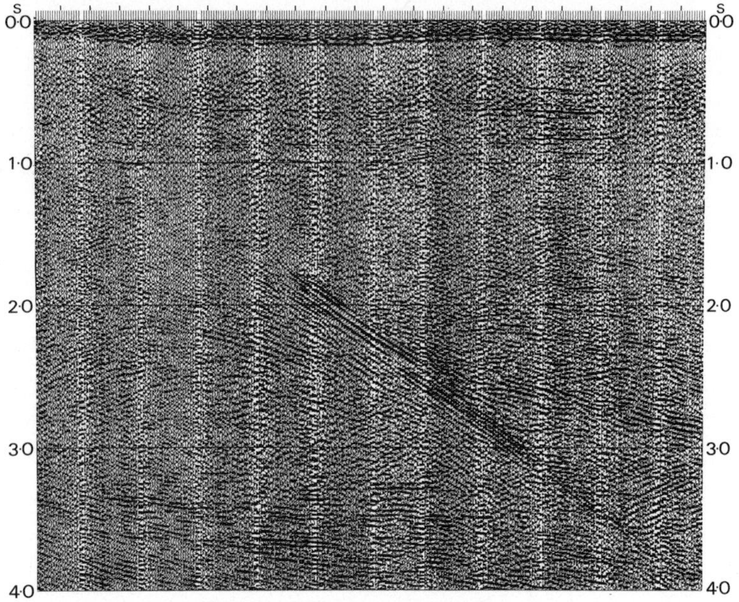

Fig. 41. A section corresponding to the sketch of fig. 39.
This section has rather a poor signal to noise ratio, and so should be viewed from the right-hand edge.

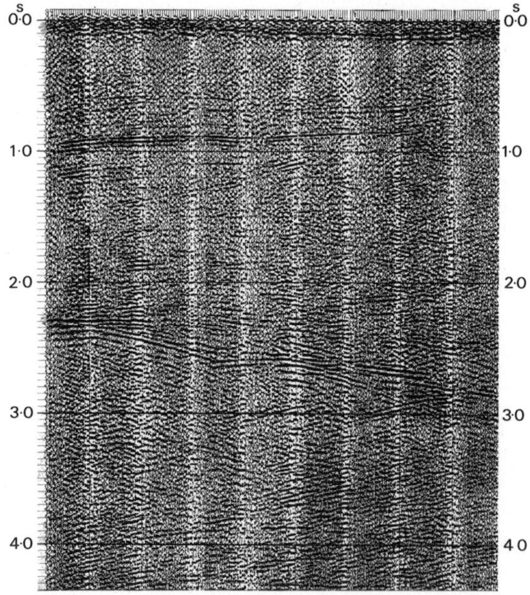

Fig. 42. A section corresponding to the sketch of fig. 40.

Figs. 34 and 35 show the seismic response of an asymmetrically folded series overlain by a nearly horizontal sequence. Each steep limb has a 'shadow area', or unexamined portion. The seismic expression shows the two straight line extensions. These two sketches are taken from real data, which has good velocity control to establish the position of the high velocity bed.

An example in which the high velocity beds are involved in the folding is shown in figs. 36, 37 and 38. Fig. 36 shows the data from a traverse across a strong fold, developed in isolation in a flat or monoclinal area. Fig. 37 sketches the elements of the arrivals from a single horizon, showing the reflections from the flat and monoclinal areas on either side of the fold, the partial anticlinal reflection, and the inverted synclinal expression – also a partial feature. The sketch also shows the refracted reflections which continue these reflections. Fig. 38 shows a first attempt at constructing a model which would, qualitatively, give rise to the observed arrivals. A further examination shows it to be more likely that the steep limb of the fold is overturned.

Next the sedimentary complexities are illustrated by a great reef development. They are so vast that they have rarely been bottomed by drilling. There is, in fact, one record of 3000 m of reef limestone in a hole which bottomed within the reef.

Both fore-reef and back-reef faces are rather steep. The off-reef sediments, and the overlying sediments are flat-lying, and contain rather high velocity beds, though not velocity inversions as far as is known.

Rays from an observation point close to the reef, at the correct angle of obliquity, strike one of these refracting faces at the critical angle, travel along the refractor to the reef, where they are reflected back and refracted back to the observation point. The relationships are shown diagrammatically in figs. 39 and 40, and some real data is shown in figs. 41 and 42.

It will be noted that the reef front has a somewhat fluted form, so that the reflected refractions seen on the seismic section parallel with the front of the reef trace a series of hyperbolas. There are several refractors, giving rise to a series of interfering shapes. If the refractors were nearly horizontal, as in this case, then each shape is roughly a contour on the reef face. Its depth can be determined by relating the refractions to the corresponding reflections in the off-reef sediments, and making a time-depth conversion.

9.6 Quantitative machine migration v. qualitative interpretation by inspection

It will be seen from the foregoing that the interpreter can, by inspection of the cross-section, recognize features and interpret them. Sharp synclines, overturned anticlines, faults, and other complexities on the seismic section can be transformed, in the mind's eye, to the corresponding real earth features. One cannot, however, derive the precise shapes by inspection, and for some features the precise position cannot be specified.

By contrast, the computer routines for migration will show the reflections in their correct earth positions if the problems of migration velocity and time-depth velocities have been solved sufficiently closely. But the continuity on such migrated sections will be broken where no reflection has been recorded, in the 'shadow zones' of the previous section. The 'shadow zone' is effectively increased if the scan of the data in the migration process is too narrow to deal with the steep dips recorded on the margin of the 'shadow'. The unusual seismic arrivals on a complex seismic section include refractions, refracted reflections, reflected refractions, diffractions from point sources outside the plane of the section, and some others. All of these are treated as if they were reflection arrivals, and migrated accordingly. The outputs from these alignments on to the migrated section are confusing and misleading.

Thus the interpreter, working on the unmigrated section, will be qualitatively correct: he will have identified the features, but he is left with error in his attempt to draw their real earth shapes, and sometimes, the position. Working on the migrated section, he will delineate with much greater precision those features which are correctly identified: but there is a much greater risk of "blunder" – of mistaking the 'migrated' version of some other alignment for the continuation of a reflection. In both cases, of course, the 'shadow zones' are breaks in continuity: but these are easier to recognize and reconstruct on the unmigrated section.

The conclusion will be that the migrated section should not be interpreted without reference to the unmigrated one. A further conclusion is that difficult and complex sections will require the further stage of modelling.

9.7 Representation of complex structures seen on unmigrated sections

A recurring task is that of presenting, on maps and sections, what the interpreter has seen on complex seismic sections. Isochrons are misleading, since the sharply folded synclines will appear convex upwards, and the faults will be misplaced. No rigid solution is possible without migration.

There is a technique which is helpful for reconnaissance interpretations, particularly in the field. One can accept points on the axes of synclines and anticlines as being close to true positions. Points of inflection on the flanks can be migrated by hand. Similarly, reflection points on either side of a fault can be migrated by hand calculation. With these established points as guidance, the structure can be sketched in on cross-section, and contoured.

9.8 Representation of faults from migrated and unmigrated sections

First the seismic expression of faults will be examined rather more closely than in 2.4 of Level 1.

There are two points which are almost identical in the real earth, which may appear widely separated on the seismic section. They are the point on a reflector at its termination by a fault, and the corresponding point on the fault itself.

The reflector point generates a last reflection point in the sequence of reflection continuity, and also a hyperbolic diffraction arrival from the point of termination. The fault point generates a point on a direct fault plane arrival, which may be a simple reflection if the fault plane is not too steep; or the arrival may involve refraction also at some level.

All of these are brought to the same point by migration. The migrated point corresponds with the vertex of the hyperbola.

Along the trace of a typical fault plane, some beds will be brought against others with velocities differing in the same sense as that of the beds overlying them. In this case, following Trorey, the forward hyperbola on the seismic section – the one that crosses into the opposite fault block – will have the same phase as the reflection which is terminated. In the other case, where the bed is brought against one whose velocity differs in the sense opposite to that of the overlying bed, there is a reversal of phase between the reflection and the forward hyperbola which is a continuation of it.

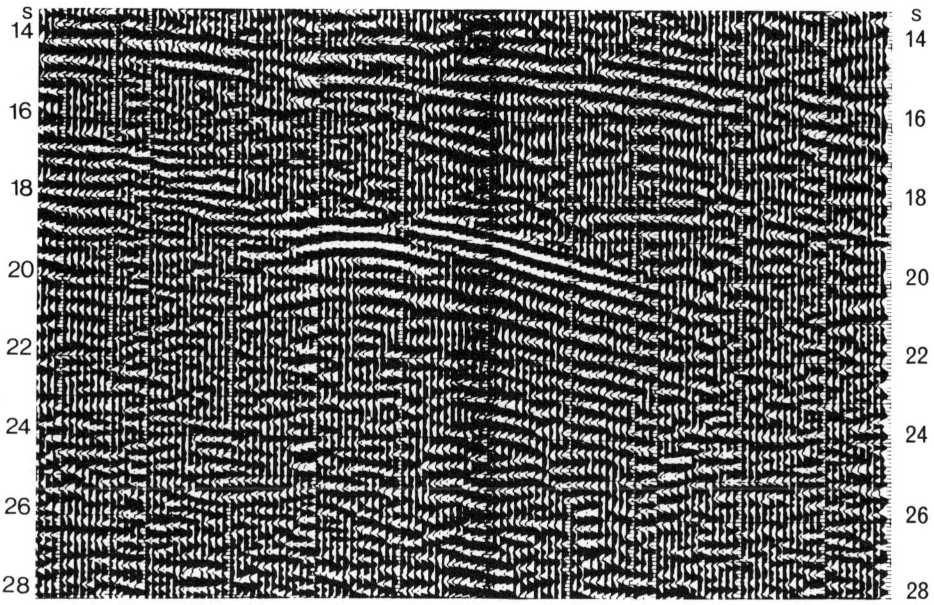

Fig. 43. Near the middle of this section are shown reflections between 1.8 and 2.0 which are truncated. These right-dipping reflections pass over into hyperbolas which are tangential to them, with a clearly defined phase change at several horizons.

Fig. 43 is an example of this change of phase. It is not always seen.

On statistical grounds one would expect the phase reversals to be more plentiful from the upthrown side of a fault but this has not been demonstrated.

Precise picking of phase reversals gives a seismic position for the fault: but to position it correctly the points must be migrated with the dip of the reflector at termination – not the dip of the fault plane reflection, if there is one. The phase reversals result simply from the mis-match between the two sides of the fault, or sometimes from the phase change of hyperbolas. The cause of the reversal can rarely be identified with certainty.

A flexure, from its geometry, cannot produce reversals of phase of this kind: and the forward hyperbolas should always be of the same phase. This appears to be the only property which can be used to interpret the seismic section which shows a broad band of confusion, limited by two lines of vertices of hyperbolas. If the forward hyperbolas sometimes change phase, the feature is probably a broad fault zone, filled with rock fragments of other material. If none of the forward hyperbolas change sign, the feature is a flexure. A second theoretical feature is that with a broad fault zone the curvature of the hyperbolas should be the same on both sides of the fault at a given time. If the feature is a flexure, the curvature should be greater on the synclinal side than on the anticlinal side. Both of these criteria are difficult to apply in practice, though fig. 20 is an example where the properties of a flexure can be observed.

Precision in delineation of the termination of beds at a fault is important when preparing isochron maps. The trace of this termination should be located accurately on the map. This trace will shift in position from one horizon to another unless the fault is vertical.

Precise plotting must take account of the observation that the terminations of a reflection on the two sides of the fault will not be coincident in plan. If the fault zone is narrow, there will be a gap across a normal fault, and an overlap if the fault is reversed. If the fault zone is wide, both types of fault may show a gap. Marginal notations should be used to record the fault characteristics on the map.

Migration tends to obscure a fault zone by artifically extending the reflections a short way into it. Migration clarifies the fault relationships by shrinking the hyperbolas to points or nearly so. Any fault plane reflection is also migrated back to the migrated fault trace, or nearly so.

The net result is to clarify the reflection alignments by migrating away the other arrivals, and to extend them up to, or almost to, the migrated fault trace. The appearance is now the butt-jointed appearance of wood-work. The reflection alignments are about in the same direction, but they do not match across the fault trace.

9.9 Synthetic studies

Enough has been written to show that seismic interpretation is not a matter of rigorous deduction, so that the interpretation can rank as a 'fact'. The data is usually imperfect in some degree, so that interpreters may differ in the alignments which they select. Model experiments have demonstrated that the seismic response of a simple geological model may be more complex than one expected. The impression is left that there remains some subjective element in seismisc interpretation: and

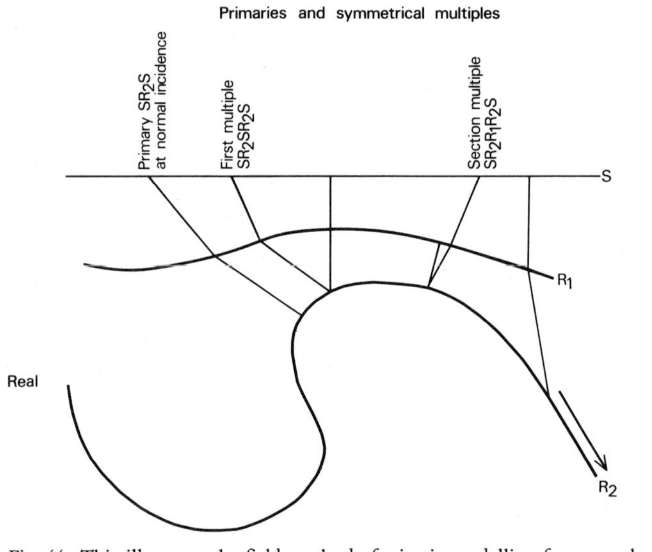

Fig. 44. This illustrates the field method of seismic modelling from a real earth structure by drawing normals to the reflecting face, tracing them to the surface, timing the path and doubling it, and then plotting the reflection time vertically below the observation point. The observation point is the surface point at which the traced ray emerges.
Symmetrical multiples can similarly be modelled, by starting at the normal incidence surface, (the point of symmetry for the path), and travelling up or down initially as the path requires.
Asymmetrical multiples cannot be treated in this simple way, since they have no normal incidence on any surface.

this is confirmed by finding differences in the interpretation of the same data by different workers. Usually these differences are minor ones.

Larger differences appear when seismic interpretations are compared with the results of other attempts to portray the subsurface geology using entirely different data. Examples are interpretations of gravity data, and subsurface geological maps which often require interpolation or extrapolation over wide distances without well control.

The seismic interpreter can reduce these different interpretations to a common basis by reconstructing the seismic response of alternative interpretations. First cross-sections are constructed along the lines of traverse: or in the case of sections oblique to the dip direction, normal incidence sections.

Fig. 44 illustrates the method of construction of a synthetic section. Points equally spaced along a reflector are marked off. Then, making use of the known interval velocities, rays normal to the reflector are traced to the surface and timed. Where the surface points which result are widely spaced over an anticline, additional points are added on the reflector. Truncations of reflectors at faults, and fault plane reflections, can be similarly dealt with. If the normal to the reflector fails to get to the surface, a shadow zone is established.

Symmetrical multiples can similarly be traced and timed by starting at the reflector which marks the point of symmetry, setting out normal to it in the correct sense, and tracing the ray through its successive reflections to the surface. A few such multiple paths are illustrated.

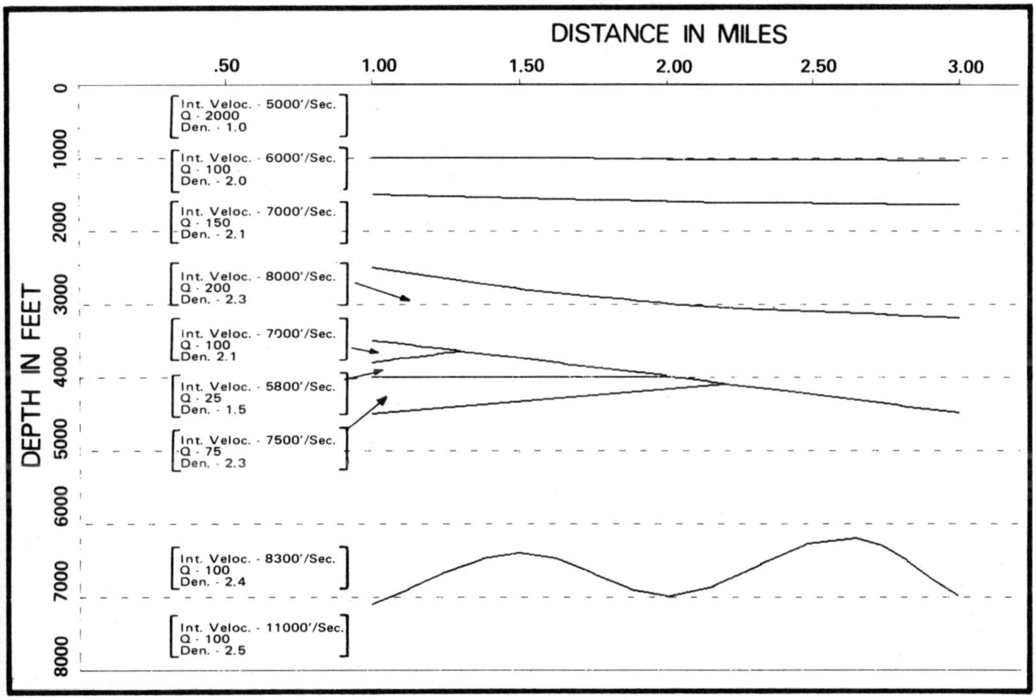

Fig. 45. Earth model showing reflecting interfaces with velocities, densities and absorption/dispersion factor indicated.

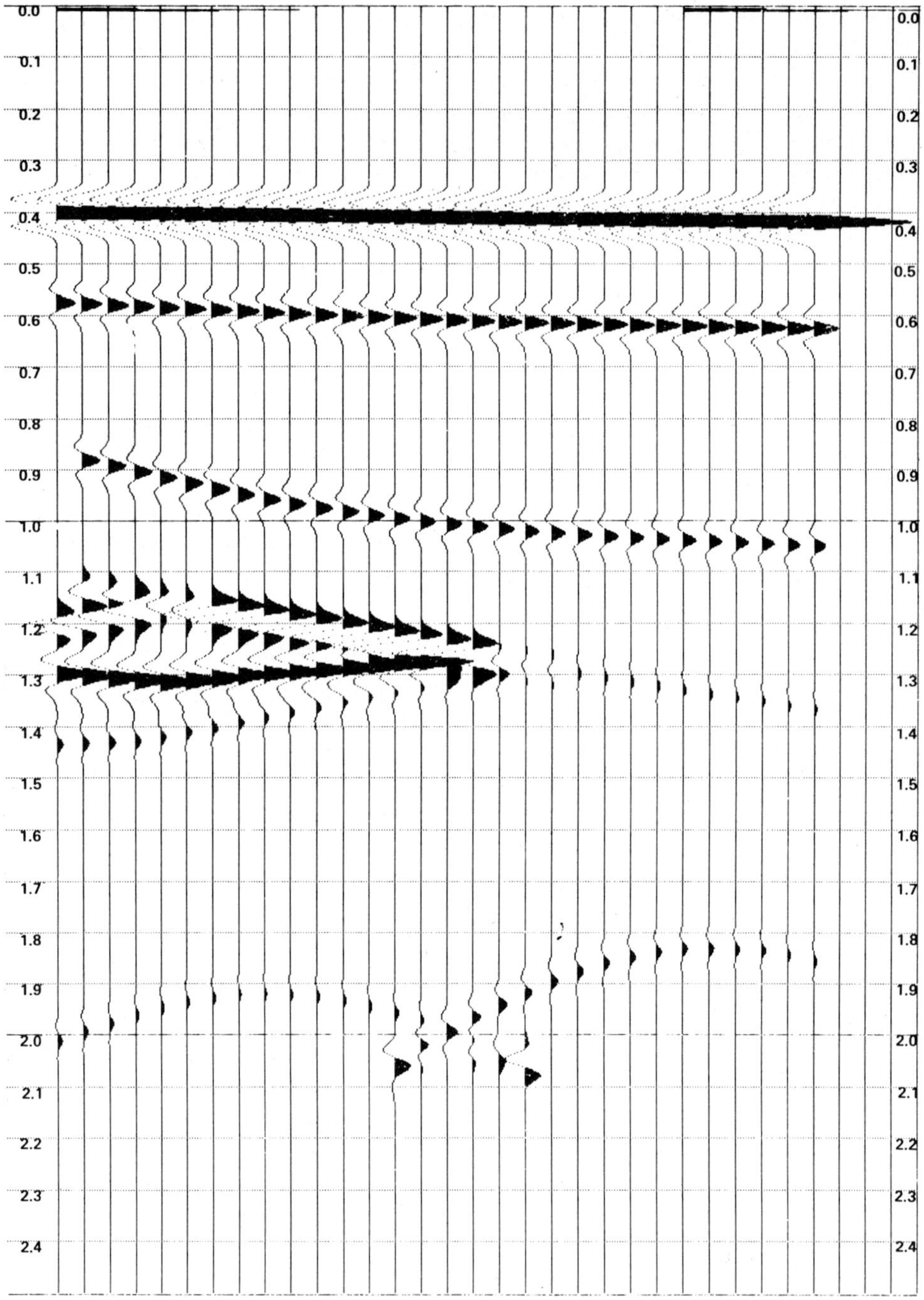

Fig. 46. Synthetic spike seismic profile convolved with RICKER wavelet.

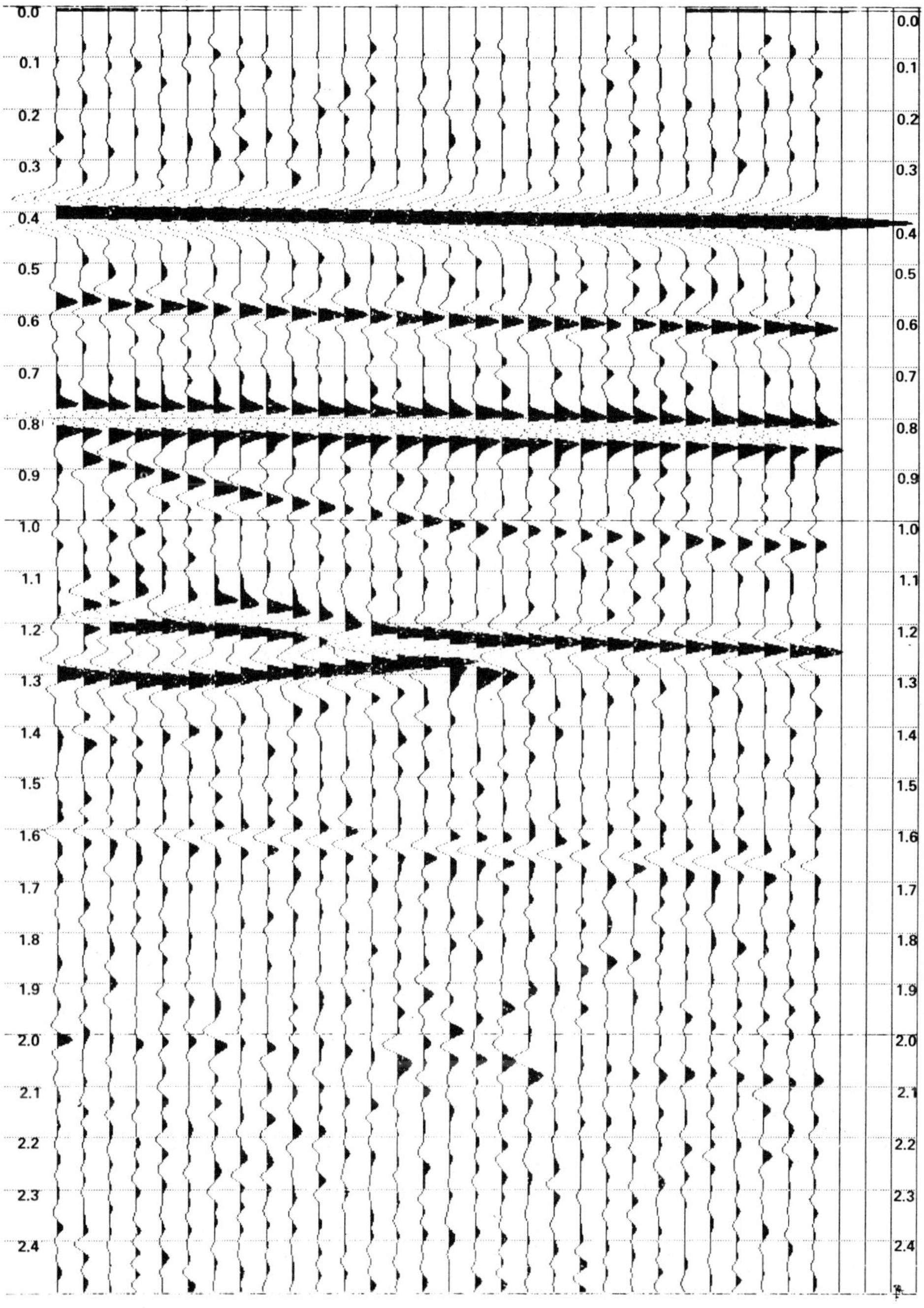

Fig. 47. Synthetic RICKER seismic profile (same as fig. 46) with multiples and random noise added.

Asymmetric multiples are much more difficult to handle, since they require re-iterative calculations to establish the path which returns to the point of origin. There is no point of normal incidence from which the construction can be begun.

Some other complex arrivals can be modelled by this route – for example refracted reflections and reflected refractions.

Comparison of these synthetic sections with the recorded data will identify those features of the observations not accounted for by the model, and will indicate also model features which must be modified or discarded to fit the observations.

9.10 Computer modelling

The previous section describes a method of hand construction of synthetic sections which is slow and tedious, but is often effective under field conditions.

The nature of the construction and the calculations involved show that it is easily adapted to computer calculation, and this has been programmed.

In addition more elaborate modelling programmes have been written which synthesize not only reflection times, but also pulse form and amplitude. In later chapters will be found some discussion of reflection amplitudes. A valuable beginning has been made in the interpretation of amplitudes, though much remains to be done. Little has been done in the interpretation of pulse form.

Figs. 45 – 47 illustrate a current modelling programme. Fig. 45 is the model, fig. 46 is a presentation of the primaries, fig. 47 adds sea-bottom multiples and some random noise.

Seismic modelling, as it develops, must examine the complete seismic motion throughout the model, and trace it to the surface to approach closely to the real phenomena and earth response. This enters a field of great complexity since the wave equations must be used. It increases computer requirements since a large area or volume of the model must be stored. The paper by Kelly et al. (1974) shows one direction for these developments.

Appendix to level 2, section 9.3

Geometrical relationships for a dipping reflector

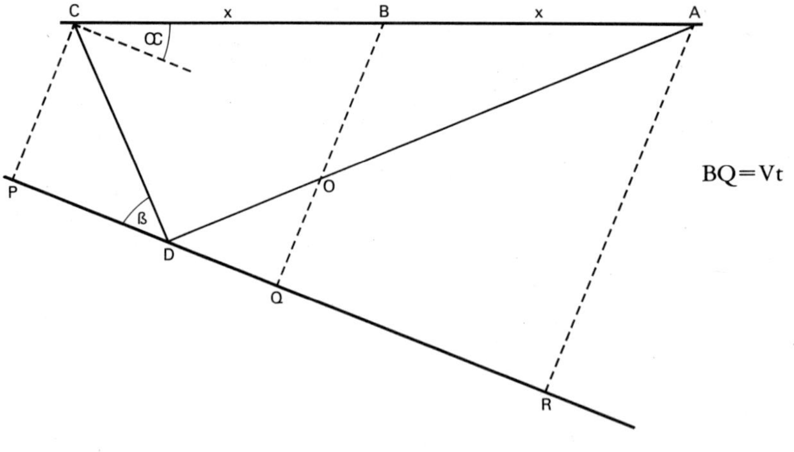

$$(AD+DC)\ \cos\beta = 2\ x \cos\alpha \qquad (I)$$
$$(AD+DC)\ \sin\beta = 2\ Vt \qquad (II)$$
$$(AD-DC)\ \sin\beta = 2\ x \sin\alpha \qquad (III)$$
$$(AD-DC)\ \cos\beta = 2\ QD \qquad (IV)$$

Combining (I) and (II),

$$\tan\beta = Vt/x\ \cos\alpha \qquad (V)$$

Combining (III) and (IV),

$$\tan\beta = x \sin\alpha / QD \qquad (VI)$$

Equating (V) and (VI),

$$QD = \frac{x^2 \sin\alpha \cos\alpha}{Vt}$$

or

$$QD = \frac{x^2 \sin 2\alpha}{2\ Vt} \qquad (VII)$$

Some interesting conclusions arising from the geometry are:

(1) A line through point P, parallel to the surface, passes through point 0.

(2) O is the 'half-way' travel time point on the path AODC.

(3) An ellipse, centre at B and foci at C and A, passes through the point D. The reflector is tangential to the ellipse at D.

(4) Equation (VII) implies that QD/OQ=PQ/BQ
PB is / / to DA and BR is / / CD.

(5) A rectangular hyperbola, origin at B and one axis along the line BQ, passes through the points D and C and A. The Hyperbola 'constant' is $x^2 \sin\alpha \cos\alpha$.

(6) Combining equations (I) and (II) in a different manner $(AD+DC)^2 = 4x^2 \cos^2\alpha + 4V^2 t^2$ and oblique path travel time is.

$$2\left[t^2 + \frac{x^2}{V^2}\cos^2\alpha\right]^{1/2}$$

Some of these relationships find practical applications. For example (3) above is the basis of migration before stack. (6) shows that the stacking velocity for a dipping layer is $V/\cos\alpha$.

10 The precision of timing

10.1 Standard error of reflection times

It is curious that in a time of rapid advance in all of the aspects of seismic exploration there should be some small aspect of the work where performance has deteriorated. Older geophysicists will recall picking wiggle traces, one by one, to establish continuity. They will recall, also, careful timing of reflections, striving for the nearest millisecond of reading, so that the reading error was generally a half millisecond or less. Error from other sources was often more than this, but, at least, the reading error did not make a major contribution to the total error.

In the present day the timing accuracy of the equipment used is rather better. Yet many interpreters work with small scale sections, and their continuity is picked to the nearest 5 ms or so, with occasional demonstrable errors of 10 ms. The reading error is now about 10 times what it used to be.

There is little on determination of the standard error of reflection times in the literature. Some observations were made by the writer on a good reflection: the base Tertiary reflection in the southern part of the North Sea. Here repeat surveys of the same line of traverse yielded statistics on the scatter of the data. This included all the operational sources of error; including, in this case, any errors due to position errors for the later runs.

The standard error determined was 1½ ms, which includes the reading error. It is concluded that the standard error of the data was a little above 1 ms, and that 1 ms represents about the best value likely to be achieved in practice with good reflections. This determination was made on single cover data, as recorded.

At the other limit, it will be apparent that if the standard error of a reflection time exceeds 10 ms, the alignment will be very erratic, and the record will be approaching the state where continuity, and the reflection itself are unrecognizable. Most of the interpreter's work is carried out with material having a standard error between 1 and 10 ms.

Next one must consider the effects of high mulitplicity of recording. 48-fold stack should reduce the standard error due to random variations to about one-seventh of its single cover value. Other data processing procedures still further improve the signal to noise ratio, and hence the standard error. Finally, the geometrical constraints which limit the nature of reflection time change from trace to trace justify certain smoothing functions, which still further reduce the standard error of individual reflection times.

The conclusion to be drawn is that the best recording and processing procedures, applied to good reflections, yield reflection times whose standard error is of the order of 0.1 ms. There is some indirect support for this. Interval velocity determinations between good reflectors show rather less scatter if the reflection times are interpolated to 0.1 ms than if they are used to the nearest sample.

It is probable that the inherent precision of the seismic reflection is not fully exploited, but there is not enough evidence to justify the case fully.

10.2 Time-break

Precision of timing requires an equal precision in the establishment of the zero from which the times are measured. There is a wide range of methods for recording this 'time-break' or time-zero. Most are designed and tested to 1 ms accuracy, which is an accepted standard for technical work. There must remain a suspicion that the 1 ms standard error of good reflections has, as a major contribution, a time-break with a standard error of almost this size. This has not been demonstrated.

10.3 The 'true zero' of time

One can envisage ways in which 'time-break' registration can be improved to give higher accuracy. But one must then turn to the physical and analytical significance of the time-zero.

In the immediate neighbourhood of any seismic source there is a high density of energy in the wavefront – in joules per sq. m. Always this density exceeds the elastic limit of the materials traversed by the wavefront. The distance for which this condition persists may be tens or even hundreds of metres with explosive sources, or a few centimetres only with Vibroseis. Non-linear propagation takes place in this zone in which the elastic limit is exceeded, usually as shock-waves of very high velocity. At some nearly spherical boundary the particle acceleration just fails to exceed the elastic limit, and from here on the classical seismic relationships apply.

A useful time-zero would be that arrived at by extrapolating from this spherical boundary backwards to the theoretical point source of calculations, using the appropriate seismic velocity. A number of phenomena would then appear simpler. For example, sea-bottom reverberations would extrapolate backwards in time to a zero member at such a time-zero.

Older seismic work, with its more accurate timing, showed an awareness of this. Successive shots from the same hole showed a time-lag, a measure of the increasing diameter of the fractured, low velocity zone created by the successive shots. A heavy charge, following an initial small one, showed a lead, probably because of the longer shock-wave path.

10.4 The concept of reflection time

Contemplation of higher precision of reflection times leads to a re-examination of other seismic quantities besides the time-zero. Velocities, NMO corrections, statics, all take on new aspects when 0.1 ms is considered.

The relation of the arriving reflection pulse to the theoretical spike position must also be examined.

Two quite different influences are at work.

Absorption in the earth diminishes the high frequencies more than the lows, and causes the pulse to spread in time.

Transmission through a layered earth transfers the energy from this initial broadening pulse to the first lobe of the reflection tail as the pulse advances. Eventually, the initial pulse is insignificant, and is lost: moreover, the contribution from its peg-leg multiples to the first lobe becomes feebler and feebler, and for practical purposes the first lobe becomes the seismic pulse. It will be clear that this, in its turn, will now lose energy to the second lobe with still further transmission. This transference of energy, and of recognizable amplitude to later and later lobes is familiar (though with a quite different model) in tracing refraction arrivals to increasing distances.

This change is a progressive one in nature. The change is accentuated by the use of time-variant filters. On some sections it is possible to recognize very high frequency, spiky reflections at shallow depth, reflections of intermediate width at moderate depths, and low frequency reflections at great depth. It is interesting to speculate that this represents successive stages of transfer of amplitude to later lobes: but it has not been demonstrated.

11 Precision of velocity determination

11.1 The precision of interval velocity

From the usual equation, the r. m. s. velocity $\overline{V}$ is given by

$$T_x^2 - T_0^2 = \frac{X^2}{\overline{V}^2}$$

It can be derived, that if δT_x, δT_0 are the small errors in a measurement of T_x and T_0 respectively, then the resulting small error $\delta\overline{V}$ in $\overline{V}$ is approximately given by

$$\delta\overline{V} = \frac{-\overline{V}\,(T_x\,\delta T_x - T_0\,\delta T_0)}{T_x^2 - T_0^2}$$

Moreover if δT_x, δT_0 are independent random errors, each distributed with zero mean and standard error σ, then $\delta\overline{V}$ is randomly distributed with zero mean and standard error $\sigma(\overline{V})$ given by

$$\sigma(\overline{V}) = \frac{\overline{V}\,(T_x^2 + T_0^2)^{1/2}\,\sigma}{T_x^2 - T_0^2}$$

i. e. the standard error $\sigma(\overline{V})$ increases with time, and with decreasing step-out, as is to be expected.

Example $T_x = 1.2$ $\quad T_0 = 1.0$
$\overline{V} = 3000$ m/s

Then for $\sigma = 0.001$ $\quad \sigma(\overline{V}) = 11$ m/s
$\sigma = 0.010$ $\quad \sigma(\overline{V}) = 107$ m/s

It will be apparent that if the standard error of reflection time can be reduced to 0.0001 sec, then the velocity standard error will be of the order of a metre per second in this case. Interval velocities of geological intervals can be determined to a few metres per second if they are not too thin. Velocities of this precision are valuable aids to the interpreter: but before discussing the ways in which they can be used, it is necessary to look at the weaknesses in the process as discussed so far.

One-tenth of a millisecond of reflection time is generally less than 0.3 m of depth in the subsurface. Working to this level of precision requires a re-examination of the detailed match between micro-stratigraphy and pulse. This is often helped by synthetic studies based on velocity and density logs.

The step-out values, and hence the values for velocity derived from them by standard formulas, are affected by dip and by curvature. The effects are negligible for small values, but for strong dips and curvatures the results can be useless unless they are corrected for these effects; and even with good control the error is still increased.

Thus the statistical error based on a simple model is not much guide to what the results would be in a complex earth. A broad conclusion is that under fairly simple geological conditions one can obtain results with a small absolute error and a small scatter. As the conditions become more complex, the absolute error increases (because the model is less close to the real earth), but the scatter is not notably increased. Thus those velocity uses which require absolute accuracy are restricted to simple areas: but those based on relative values have a wider application.

11.2 Use of velocity by the geologist

Interval velocity is a rock property. It is the only property which can be measured over a relatively small block of material in the subsurface without significant effects from surrounding materials: and it can be measured from the surface.

To a limited extent the absolute value can be used for lithological identification, but almost always in the sense of excluding certain possibilities. There is too much overlap of the seismic velocities of different rock types for most of the range to make positive determinations. Velocities are of greater value in relating a sequence known by its seismic velocities with those in a well or at outcrop. A special use of velocities, with other data, is in the recognition of change of fluid content, as from water to gas: or the recognition of facies change, as from sand to shale, and mapping these features in the subsurface.

11.3 Use of velocities by the data processor

Velocity values enter into the work of data processing in three principal ways:

Calculation of normal move-out and other corrections.

Migration.

Time-depth conversion.

These velocity values, expressed as a function of T, the reflection time, are about the same, but they diverge under extreme conditions.

The use of 'velocity' in connection with NMO corrections is rather misleading. The procedure adopted is an experimental determination of the step-out values to be applied to a common-depth point gather to obtain the maximum reflection output on stacking. This experimentally determined value is the one to apply.

It is often helpful to determine the 'stacking velocity', i.e. the uniform velocity above the reflector, or the r. m. s. velocity in a horizontally layered model, which would produce the same step-out pattern. These 'stacking velocities' contain all the effects of dip, curvature, and changes of pulse form, but this is quite immaterial for the purpose of stacking.

It has been seen in the appendix to 9.3 that the reflecting point moves up dip with increasing source-detector distance, and that the velocity derived on the assumption of a horizontally layered model is V sec α, where V is the true velocity and α the angle of dip. Thus the error is 0.1% with a dip of 2½°, 1% with a dip of 8 °, 2% with a dip of 11°, and so on. It reaches 10% with a dip of 25°.

Migration, in which the search point remains fixed, gives a step-out for the wider search angles which is larger than that in the CDP model, and consequently requires a velocity which is much closer to the r. m. s. velocity derived from the rock properties. Broadly, the stacking velocities are much more affected by the structural developments than are the migration velocities.

Finally, the average velocity, which is applied to the stacked and migrated section to convert it to depth is the weighted average velocity. Formally, to a fair approximation

Stacking velocity is given by

$$\bar{V}^2 = \frac{\sum_{i=1}^{n} v_i^2 t_i}{\cos^2 \alpha \cdot \sum_{i=1}^{n} t_i}$$

Migration velocity by

$$\bar{V}_m{}^2 = \frac{\sum_{i=1}^{n} v_i^2 t_i}{\sum_{i=1}^{n} t_i}$$

Time-depth conversion velocity by

$$V_a = \frac{\sum_{i=1}^{n} v_i\, t_i}{\sum_{i=1}^{n} t_i}$$

where the v_i are the measured rock properties.

Some interesting relationships between velocity and migration are discussed by GARDNER, FRENCH & MATZUK (1973) and FRENCH (1974) in the papers already cited.

11.4 Use of velocities by the interpreter

The significance of interval velocities for the interpreter is that they afford him a separate approach to seismic interpretation. A plot of interval velocities over the area of a cross-section will enable him to draw in lines of equal interval velocity, which approximate to reflection continuity. Across any gap in the data he can establish correlation by using the velocity column on either side. The velocity interpretation should, of course, confirm the interpretation of the reflections. Any differences are important and deserve special study.

11.5 A common ground for discussion

One important task in exploration is to ensure that all of those working on a project have the same model of the earth. This integration of effort is one in which the interpreter should take the lead. A difficulty is always to find some common ground on which to meet.

Interval velocities often provide a starting point: geologist, data processor, and interpreter, all understand and use this rock property. It is a good start to the process of integration to ensure that all of these workers have the same model.

12 Interpretation of amplitude and pulse form

12.1 Introduction

Amplitude and pulse form have long been the elements which the interpreter surveys as he searches the seismic section to establish continuity and correlation.

The special interest is the newer technique of interpreting amplitudes in terms of fluid content. The success of this has led to very detailed studies of amplitude interpretation, along with associated velocities, phase of the reflection and other data.

Some of these later studies are difficult to assess in terms of practical usefulness. They can perhaps be put into perspective by saying that there are many effects which influence the amplitude of

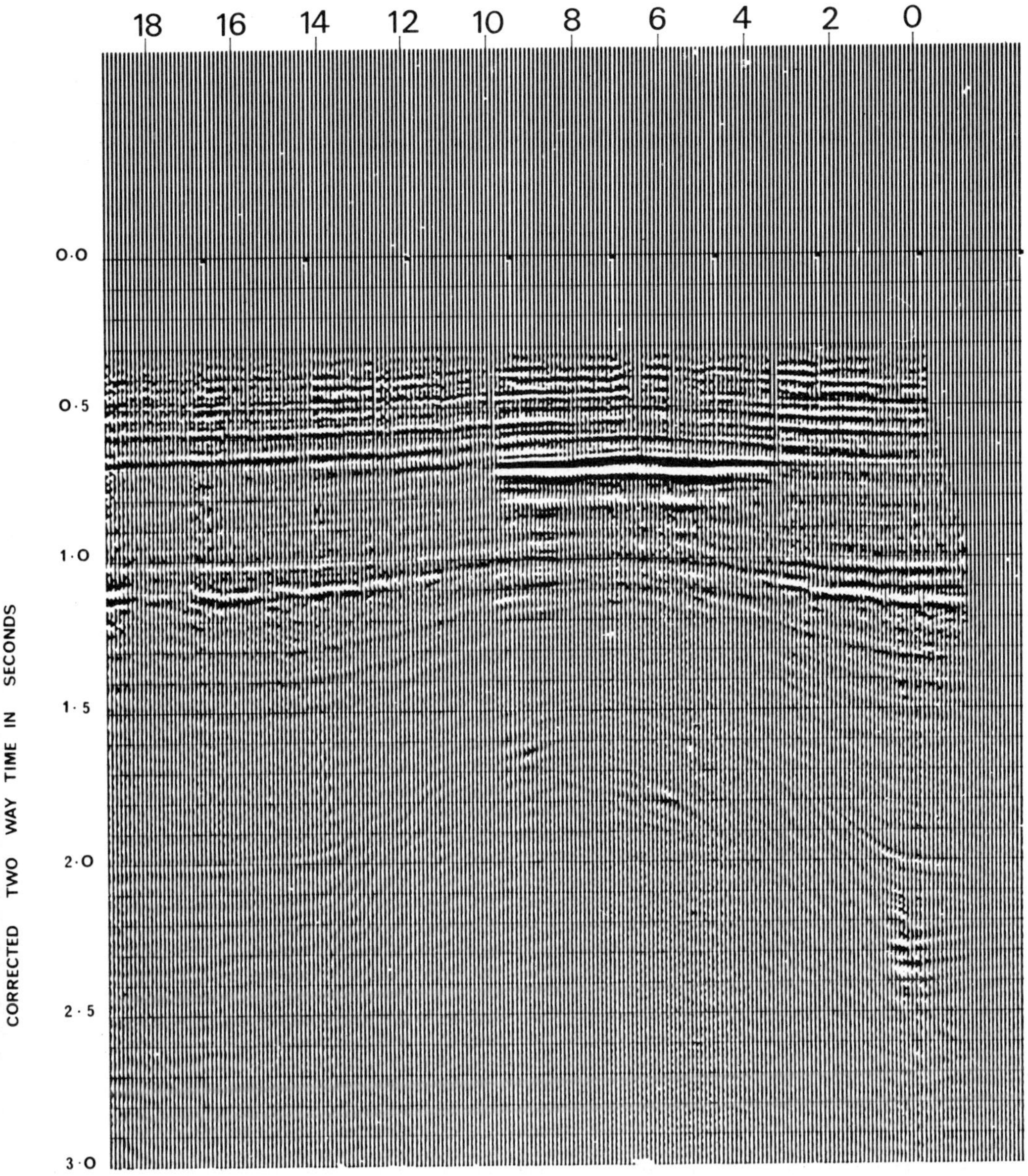

Fig. 48. A cross-section with true amplitude processing.
Note the very high amplitudes associated with the gas sand at the crest of the structure. The amplitudes are so high that residuals from multiples of the gas sand reflections are still high enough to obscure low amplitude primaries.
Gas pockets such as this are more important as hazards for the driller than as potentially commercial gas sources, unless they are very favourably situated.

a reflection. Most of these can be calculated, and the changes of amplitude derived. The observed amplitude is the sum of several or all of these influences. The difficult task is to find how much amplitude is to be assigned to each influence: if this can be done, then the reverse calculations, leading to quantitative interpretation, can be carried out. Rarely can this assignment be done with any precision; though it is often possible to eliminate a number of the effects as having no significant contribution to amplitude.

The principal success with amplitude interpretation has been in the interpretation and recognition of gas sands. In this case the amplitude changes from this cause are very large, larger than those observed from any other source, and the velocity data provides strong support. Moreover, the geological environment in which the gas sand appears is usually well-defined, and recognizable on the seismic section. Fig. 48 is an example.

The paper by R. E. SHERIFF "Factors affecting amplitudes – a review of physical principles" (1973), is important reading on this subject. It will encourage caution in the interpretation of amplitude data.

12.2 Factors affecting amplitudes

The factors affecting amplitude have been well summarized by SHERIFF and this section is based on his writing. The headings of his discussion are:

1. Source strength and coupling.
2. Spherical divergence and ray path curvature.
3. Absorption.
4. Scattering.
5. Peg-leg multiples from thin reflectors.
6. Superimposed noises of various types.
7. Interference of different events.
8. Wavefront healing by diffraction.
9. Attenuation in the near surface.
10. Geophone sensitivity and coupling.
11. Array directivity.
12. Transmissivity.
13. Instrumentation.
14. Reflection at interfaces.
15. Reflector curvature and velocity focussing.
16. Reflector rugosity.
17. Variation of reflection coefficient with incident angle.

Of this long list, only the last four contain information about the reflecting interface, and only 14 contains the factor on which current amplitude interpretation is based. 2, 3 and 5 contain subsurface information of another kind.

It is of interest to note, also, that of all these factors only 2 is essentially independent of frequency. Only one factor, in other words, has no influence on pulse form.

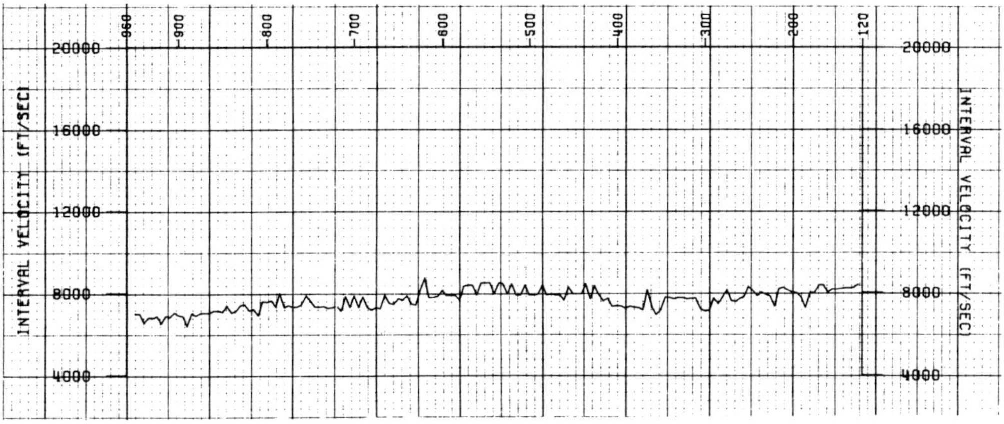

Fig. 49. Interval velocity of the bed above the gas sand of fig. 48.

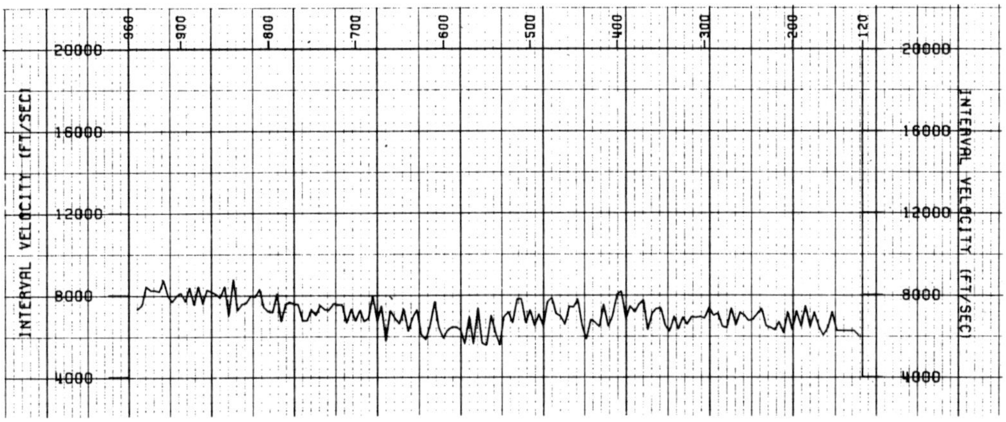

Fig. 50. Interval velocity of the gas sand of fig. 48. Note the velocity change from about 8000 to 7000 f.p.s. as the gas is approached from the left.

From this summary it will be apparent that an interpretation of amplitudes can only be valid if all of the factors under control have been standardized or normalized: and all those not under control have been assessed and shown to be negligible, or else assessed and subtracted from the observed amplitudes.

12.3 Gas sands

Seismically, a gas sand is characterized by a low velocity and a low density. The changes are often of the order of 30% in velocity and 20% in density when compared with the same sand water-saturated.

With changes of this order there is no limit to the possible amplitude changes. Changes of the order of ten times can be observed on sections. Such indications are very striking on sections which

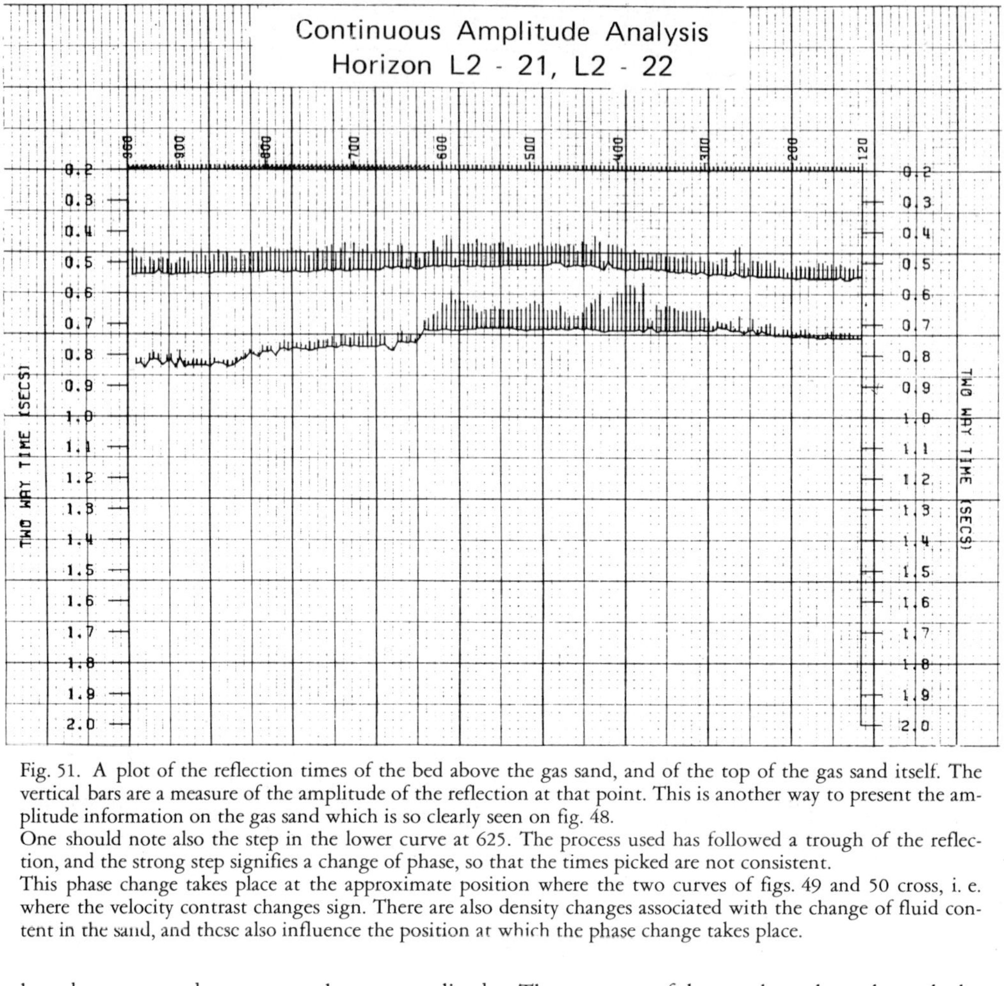

Fig. 51. A plot of the reflection times of the bed above the gas sand, and of the top of the gas sand itself. The vertical bars are a measure of the amplitude of the reflection at that point. This is another way to present the amplitude information on the gas sand which is so clearly seen on fig. 48.
One should note also the step in the lower curve at 625. The process used has followed a trough of the reflection, and the strong step signifies a change of phase, so that the times picked are not consistent.
This phase change takes place at the approximate position where the two curves of figs. 49 and 50 cross, i. e. where the velocity contrast changes sign. There are also density changes associated with the change of fluid content in the sand, and these also influence the position at which the phase change takes place.

have been prepared to preserve the true amplitudes. They are powerful enough to show through the attempts to normalize amplitudes in regular processing of seismic sections.

The supporting evidence includes the change in velocity, already referred to. This cannot be measured if the bed is too thin. Even in this case, however, if the gas sand forms part of a larger thickness of beds whose velocity is being measured, the contribution of the lowered velocity will be observed.

It will sometimes happen that although the velocity of a gas sand is much below that of the shales above and below, the same sand, water-saturated, has a rather higher velocity than those shales. Where this happens, there is a reversal of phase close to the edge-water zone. This is illustrated in figs. 49 – 51.

In a well-developed gas-cap there is other supporting evidence – other manifestations of the basic physical conditions. For example, the low velocity of the gas sand may often produce a recognizable decreased curvature of the lower surface, compared with the upper. The very high reflectivity of the gas sand interfaces may produce a shadow-zone below it in which the primaries are much reduced in amplitude. At the same time, multiples involving the gas sand boundaries and the near surface become much stronger, and may obscure these weaker primaries.

Of particular value in establishing a probable gas-cap is the reflection from the horizontal gas-water or gas-oil interface, if the configuration in the subsurface, and the sand thickness, is sufficient to allow it to develop. Because of the irregular distribution of velocities in the neighbourhood of the gas-cap, the horizontal interface in the earth may not yield a 'horizontal' reflection on the seismic section. In a typical case the conflicting alignments will be clear, but in a confused case it may be necessary to test the interpretation by a seismic modelling procedure.

All seismic interpretation is conducted with an awareness of what is geologically plausible. This does not mean substituting a pre-conceived geological concept for the observed geophysical data. It does mean that in the selective viewing or scrutiny of the displayed data one is guided first to examine the more plausible possibilities suggested by a first inspection.

This guidance is more necessary in amplitude interpretation than in some other aspects of the subject. An obvious example is a section with a strong increase of amplitudes over the anticlines and a small increase in the synclines. Geologically one is drawn to examine first the evidence for gas accumulation on the anticlines: and if the evidence is there it can soon be assembled, and the case made. The same result would be achieved by examining systematically the contributions of SHERIFF's seventeen factors affecting amplitudes in both anticlines and synclines, but pace and momentum would suffer.

12.4 Remarks on pulse form

Almost all of the factors which affect amplitude have some effect also on pulse form. These factors can rarely be expressed as operators of pure magnification with no frequency dependence, and so change of form results.

In practice, the pulse form is found to be controlled much more by the stratigraphy around the reflector than by these other factors. It is more often possible to arrive at a convincing pulse-form interpretation than a pure pulse-amplitude interpretation.

Detailed pulse-form interpretation is prevented by a lack of precise knowledge of the pulse form which corresponds to a spike input. If it were known, then convolution of the recorded pulse with the inverse of this standard pulse would produce the detailed log of reflection coefficients. This is not possible in practice.

Most effective use of pulse form is in situations where accumulation of hydrocarbons is controlled by some stratigraphic change, and there is enough subsurface exploration to establish exact stratigraphic detail of the productive and unproductive configuation, and of the nature of the change from one to the other. One must have a clear geological concept of the point in this stratigraphic change which is the effective edge condition and forms the trap. Such data comes from logging and sampling wells, velocity and density logs, and surface seismic data across the wells.

One now models, geologically and in terms of velocity logs, the typical stratigraphic change which is being investigated. Synthetic records, which match the observed data at the wells, and fill in the interpolated conditions between these observations, are now prepared. With this control seismic traverses can be run across the expected trend of the stratigraphic feature, and the trap mapped by identifying the points at which the critical pulse form on the synthetic is found in the observed data.

Such a technique is often effective where a sand splits and eventually passes into shale up-dip. The stratigraphic change must, of course, be studied in conjunction with the structure.

13 Interpolation: Contouring

13.1 Sampling

It is characteristic of seismic data that it is detailed along the line of traverse, absent elsewhere. Yet visualization of the geological model must be three-dimensional, if one is to reconstruct oil accumulation processes within the model. The spatial sampling pattern of a seismic survey is quite unsuited to three-dimensional modelling. The interpolation distance is often tens or even hundreds of times the sample interval distance on the line of traverse.

The usual treatment is to select certain key reflections as mapping horizons, to plot the observed points in time or depth on a map, and to contour these values to give a representation of shape. It will be apparent that the shape of the contours – the shape of the fold surface to be envisaged – is determined by the interpolation procedure as well as by the observed data. If the grid of observation is close, i. e. much smaller than the size of the structures being examined, there is, in practice, little range of interpretation to be expected from a range of acceptable interpolation procedures and formulas. Where the grid size is about the same as that of the structures to be mapped, the structure becomes somewhat indeterminate, and the contour shape is largely dependent on the interpolation procedure.

Finally, where the grid size is much larger than the structures examined, the structure has little expression in the contours, which are now very largely dependent on the interpolation formula adopted.

If no formal interpolation procedure is adopted, and the structure map is contoured by eye, then the subjective element in different contour presentations will generally give rise to still greater differences between contour and truth.

This subject is one of great importance for the interpreter, since exploration decisions are taken on contour maps of the structure. Examples where the difficulties have arisen in the past have been in the Southern part of the North Sea, where an area of salt domes and pillows, and related sinks, often a few kilometres in size, was explored initially with a grid spacing of 60 to 80 kilometres. A second example is to be found in Russia, where some of the earlier exploration was conducted with a very large grid. Several papers were written on the grid size in terms of the size of fields of economic interest, and the sampling spacing necessary to establish them.

With close control the interpreter is aware that he is drawing contours that are almost inevitable, and that only small departures from his solution would be geologically plausible.

With wide control the interpreter is really putting on paper a geological hypothesis which happens not to disagree with the sparse observed data. In an area of irregular control density it is easy to slide from one situation to the other. It is important for the interpreter to know how to assess his control. He must know whether he is reporting a structure established by the seismic survey, or merely reporting that the seismic survey does not yet disprove a geological hypothesis.

13.2 Interpolation and contouring

Two procedures are currently in use. The first, still generally favoured, is to contour by eye.

In effect, the interpreter first inspects the data, and creates in the mind a three-dimensional shape. He then proceeds to express this shape by contours which accord with the recorded data. If he has carried out the interpretation on cross-sections he will have formed this image as he goes along.

This process of creating a three-dimensional image must be a personal process, probably varying from one worker to another. For the writer, the first inspection of the data lifts the data points on little 'stalks' to give them relative height relationships. If the density of observation is adequate, the envelope of these 'stalks' presents a geologically significant shape, which is the basis of the contouring. If the density of observation is inadequate, this process cannot be followed. Instead, one starts with a geological hypothesis, and modifies this shape until it fits the data: this modified shape is now the basis of the contouring.

The facility for creating and carrying in the mind a three-dimensional image is greatly helped by much practice in stereoscopic viewing of air photographs and photo-theodolite pairs, and of computer-calculated stereo-pairs of various shapes. After a while extensive areas of complex topography can be remembered in the solid. Stereoscopic images of line drawings such as contours are much easier to memorize in the solid. At this stage the interpreter no longer finds any barrier to creating a three-dimensional image from his two-dimensional data.

Probably this phase of seismic interpretation would benefit from the wider use of stereo-images, particularly in the training of interpreters. Good use has been made in some other fields, as in the contouring of aeromagnetic data.

Contouring does not consist in driving the point of the pencil through the data, leaving the data-points on the appropriate side of the course as though they were obstacles in an obstacle race. This will produce a solution which does no violence to the observed data, but does not look plausible. A later effort to envisage the shape will not yield anything very convincing.

These paragraphs describe a process with a subjective element which can be large. This has led, quite logically, to the attempt to contour by automatic and mechanical means free from a subjective element. There are computer programmes available which will calculate the contours, and excellent machines which will draw them. They have not had the universal acceptance which might have been expected.

Where there is a small grid size compared with structure size, machine contouring can be acceptable. It differs little from the shapes of good hand contouring. Where the grid is less close, differences begin to appear.

The source of the trouble appears to be that the automation has not been based on the process as it existed, but on a new concept of what was to be done. Hand contouring is based on a knowledge of the shapes assumed by folded rocks, and the geologist's awareness of this helps to shape the three-dimensional image which he constructs, and the contours which he draws to portray this image. Rarely is this knowledge of the shapes of folds expressed analytically, in a form which the programmer can use to write an interpolation programme as a prelude to contouring.

A number of interpolation procedures have been used or proposed in machine contouring. All are based on the concept that there is no law governing folding – that no functional expression exists for the family of surfaces which defines the subsurface geology. It is at this point that knowledge of folding is discarded.

In its place are substituted empirical rules for interpolation which are postulated to give various classes of surface passing smoothly through the data points. These rules cannot be reconciled with the shapes of folds.

One such rule says that the interpolated value at a point is the weighted mean of certain points which surround it. The points chosen may be the nearest lying in each quadrant or octant defined by the compass points or the grid direction. The weighting may be the reciprocal of the distance to the data point or the reciprocal of its square.

Other rules set out to minimize curvature of the generated surface or of the contours which result.

An older approach is to specify a high order polynomial, or some other function with many parameters, to which the contours must conform, and to fit these to the data with a least square fit or some other measure of error.

Still other treatments put the data into the frequency domain, and effect the interpolation in that domain, where visualization of shape is even more difficult.

The geologist, particularly, will appreciate that these mathematical treatments – many of them elegant in their own field – have no foundation in the natural world.

13.3 The shapes of folds

There is an abundant literature on the stress-strain relationships in rocks. Much of this can be related, geologically, to what is observed in faulting and fracturing of rocks, and to the evidence for creep and viscous flow. It can be related to macroscopic and microscopic evidence in the field of petrofabrics. But little of the work bears on the shape that folded rocks take up in response to stress.

The best summary that one can give of the state of knowledge is that, if one considers a quarter-cycle of a fold in cross-section, from the axis to a point of inflexion, then this is, in general, very close to a sine curve. Low amplitude harmonics are present, but these cause only minor departures from the sine curve.

For many petroleum structures the trace of the axial plane on the cross-section is vertical. In this case the fold is a quarter sine curve referred to rectangular axes. In some asymmetric structures, and typically in mountain structures, the trace of the axial plane is not vertical. The fold is then referred to oblique axes, one horizontal through the point of inflexion, and the other the trace of the axial plane. With this change the generalization still applies. The fold is a sine curve, with only minor contributions from harmonics.

These relationships appear to hold for large folds of the scale explored by seismic programmes, and also for the minor folds and crumpling seen in rock exposures and hand specimens from regions of mountain building.

It must be noted that these relationships apply to real earth sections, not to seismic sections. An attempt will soon demonstrate that analytical expressions for the form of a seismic reflection from a sinusoidal reflector are quite elabroate.

Interpolation can be based on the acceptance of this observation of the sinusoidal shape of folds, and from this a system of contouring using such interpolation. It has been used experimentally: but since it is not an established technique its discussion is deferred to a later part of the book.

14 The data store

In the last chapter the discussion of contouring involved some criticism of the computerized facilities offered to the interpreter. The data store, by contrast, is a facility which seems to offer the interpreter all that he could wish for. It is not used universally, more because of the lack of facilities where they are needed, than because of any lack of appreciation by interpreters.

A seismic data store is a tape or a library of tapes on which are recorded identification for each observation point, the position data, reflection times, velocities, amplitudes, and any other observed or derived quantities which the interpeter may need for map construction.

Such a store can be up-dated with revised position data or seismic data; or by the addition of new data.

Almost any other geographically related data can be similarly stored: topographic base maps, political or land information: surface geology: vegetation: anything, in fact, which bears on the exporation programme.

Many uses are found for the data store. From it, data can be printed out for inspection and study. From the data store can be derived plot tapes to produce plots of any one of the stored quantities, or combinations of them. These plots are made of a specified map area and to a nominated scale.

Finally, the data store provides the input to further analysis programmes; for example, the study of regional velocity changes within a formation.

15 Integration

The seismic interpreter is one of a group concerned with understanding the subsurface geology, and exploiting what is found. The closest members of the group are geologists, geophysicists using methods other than seismic, and data processors. All of these workers are labouring with the same earth: for all of them the truth is the same. As investigation progresses, results converge towards this truth. At some point the results are sufficiently in agreement that all of the results can be integrated to give an earth model which is the current state of knowledge. It is on this model that exploration decisions are taken.

It would be an attractive procedure to input all of the data to some programme which would output this integrated model: but no such programme exists. What can be done is to postulate a model, and then to test the model response to seismic, gravity, magnetic and other fields of investigation; and, of course, to subject the model to a geological scrutiny to establish that it is consistent with subsurface and surface observation, and that it is geologically acceptable.

There is, thus, the task of integrating results – knowing that this cannot be accomplished rigidly or uniquely – so that a model can be produced, subjected to test, and modified as necessary to bring its response closer to observation. In some large organizations this work is carried out by a special department. In smaller organizations and field offices it usually falls to the seismic interpreter or geologist.

Few generalizations can be made to help along this process of integration. It can be said, however, that good communication, by discussion or reporting, is essential. In an earlier section the importance of the agreement on interval velocities by geologist, data processor and seismic interpreter has been stressed. This is often a good start to the working together which leads to integration. For the rest, the problems met are too diverse for generalization.

In many cases the model arrived at is dominated by the seismic contribution. It is thus of great importance in communicating the model to the exploration manager, or whoever is concerned with decision-taking about the subsurface, that the confidence level is clearly communicated. The interpreter must recognize where he can see nothing, and has no contribution to make. He must recognize, also, where the observations are sound, and are not to be ignored.

There remains the 'questionable data'. The interpreter must use this classification as sparingly as his judgement will allow. Obviously, it must not be used to screen the interpreter from the consequences of mistakes or bad judgement. In practice, the 'questionable' notation is not of much value unless the interpreter can specify what further evidence is needed to remove the question-mark or the questioned data, and if possible, how to set about obtaining that evidence. The interpreter must be aware, also, that his questioned data will inevitably be incorporated in the integrated model unless the necessary 'further evidence' appears in the course of integration.

The principal task of the seismic interpreter is clearly that of carrying out his interpretation, as described in the foregoing pages. He contributes to, or carries out, the integration of his interpretation with other data.

There is a secondary task. The interpreter is aware of the exploration objectives. He is the first, and often the only one to look at the data critically and in great detail. In the course of interpreting a current survey he will often make observations which will affect the conduct of the survey. He may note a failure to obtain data at the horizons required. He may see other possible objectives which management might wish to pursue, which require a different technique. He may see reasons to curtail or extend a survey. Any such observations must be communicated promptly.

LEVEL 3

16 Noise

16.1 Definition of signal – and hence of noise

It may be of interest to examine the interpreter's definitions of signal, and hence of noise. In some later sections the impact of these definitions on the interpreter's direction of data processing will be discussed.

Signal consists of those seismic arrivals which carry the information that the interpreter needs. These arrivals are a sequence of primary, compressional reflected pulses, which show continuity in cross-section.

Noise is any other feature of the seismic traces as shown on the seismic section.

From this rather obvious signal-definition can be derived two signal-criteria.

I. The step-out values observed on the primary compressional reflections define a curve of $\bar{V}T$, where T is the normal incidence time and $\bar{V}$ the corresponding stacking velocity, applicable to the gather. In the region below the last observation the most probable continuation of $\bar{V}T$ statistically will identify the most probable primary reflection alignments. Signals obey this $\bar{V}T$ curve within the limits of experimental error.

II. The observation of continuity on the seismic section, and velocity knowledge, leads to concepts of maximum and minimum curvature of the reflection alignment. These lead in turn to approximate limiting values of d^2T/dx^2.

This concept leads to an assessment of noise as those elements which fall outside these limiting values.

Some possible developments based on these criteria are described later.

16.2 Acceptable levels of noise

In practice, it is found that an interpreter can trace continuity with confidence if the signal to random noise ratio is 1 : 1 or larger. With smaller values the confidence decreases, and the risk of blunder increases. 1 : 1 is a good working value which everyone should strive to better. One must note, however, that some early interpretation was carried out with lower ratios, and often successfully.

It is this random noise which is covered by criterion II. above.

Aligned noise includes multiples and other types of interference which appear on cross-section. It will be apparent that a much higher signal-to-noise ratio is required with aligned noise. For example, with a 1 : 1 ratio the purely seismic conclusion is quite indeterminate – one alignment is as good as the other. If one alignment is used rather than the other, this is an expression of a hypothesis which cannot be disproved by the data, but there is no positive seismic contribution.

In practice one would like a 10 to 1 ratio of the strongest signal to be traced to aligned noise for confident interpretation.

16.3 Weakness of some current definitions of noise

The definitions of 16.1 may seem unfamiliar. They will appear more attractive after scrutiny of some definitions implied by current practice.

I. Noise is all recorded data outside the frequency range $f_1 - f_2$: both f_1 and f_2 are time variant. This seems to imply that all noise is source dependent, and that the complete solution to noise problems could be achieved by improved source design. This seems unlikely.

II. Noise is all seismic arrivals with a wavelength outside the limits $\lambda_1 - \lambda_2$. It is not time-variant or space-variant. This seems to conflict notably with I.

III. Noise is a 'tail' of standard form which follows each reflection: but differs markedly from one observation point to the next.

IV. Noise is an error in the velocity or pressure ordinates measured. This error has a symmetrical distribution about zero so that the mean of several observations (as in a CDP gather) is the best estimate of the true value.

None of these definitions has much basis in the world as it is, yet they are the definitions implicit in some standard processes. These processes work, and often work very well indeed, in the sense that they do more good than harm.

It is because of these weaknesses of definition that the pursuit of the interpreter's needs must involve some re-examination of basic concepts, and an attempt to re-build the data processing procedures on revised concepts. Such an approach may give rise to a research field of interest.

17 Weakness of some current processes

17.1 Introduction

In chapter 16 a brief inspection of noise definitions was made, and some weaknesses of definition presented. Here attention is turned to the earth models on which a number of processes are based.

The real earth is complex. The earliest attempts at data processing set up simple earth models, sometimes as a matter of computing convenience rather than as an attempt to remove some of the complexities of the real earth. In the following sections a few of these models are examined.

Again one must state that the processes derived from these models work, and sometimes work very well: but the pace of further development is slow: one must examine them to see whether they are limited by the crudity of the model. Geologists unfamiliar with data processing would be astounded by the data-processors view of the earth.

17.2 Filtering

A digital band-pass filter is a long operator, typically, which is convolved with the recorded time sequence. In geological terms, a deep reflection pulse, say from the Permian, is re-shaped by the forward part of the operator. There is added to this Permian reflection pulse a proportion of observations from overlying formations, say Mesozoic and Tertiary. The mathematical justification for this procedure would be found in the random character of the seismic arrivals from the overlying beds: but they are not random.

Thus the interpreter can follow the logic of re-forming the seismic pulse by the short, leading end of the filter operator. This may contribute improved signal-to-noise ratio by use of a matched filter: or it may contribute improved resolution by an inverse filter, and so on. He is not likely to be convinced that the sequence above the horizon which he is following is random in character – or, more precisely, that the recorded data which precedes his reflection has a spike auto-correlation.

Some current practice in devising time-variant filters is even more curious. It is usually achieved by applying two separate band-pass filters and merging the outputs over an overlap time range. Consideration will show that in the zone of merging there is no simple linear change of the pass-band with time, but the principal effect is to degrade the cut-off slope. In extreme cases of the use of this procedure the filter becomes a notch filter in the zone of merging.

17.3 Deconvolution

The model of seismic data to which deconvolution is applied is that every primary reflection has the same shape. Treatment of this situation has an elegant mathematical solution. It leads to a programme in which the input is a pulse with a tail, and the output is a pulse without one. Often the treatment is extended to include pulse-shaping, which is the identical process analytically, though very different for the eye of the interpreter.

A first weakness of this concept is that the input pulse is of constant shape. The changing geometry of multiple reflections with time and the effect of absorption ensures that the reflection pulse varies with time. In regions of complex folding and faulting the multiple relationships are so complicated that the real reflection pulse on any one trace may be far from constant.

A second weakness is apparent in the assessment of the input pulse. The assumption made is that the auto-correlation of the seismic trace is the same as the auto-correlation of the seismic pulse and its tail. Even if the tail were constant, this would not be valid. The geological sequence (expressed as the log of reflection coefficients), and noise, both contribute to the auto-correlation, and this must corrupt the assessment of the tail. An additional weakness of this assessment is that there is no unique waveform which, when auto-correlated, will yield a particular auto-correlation waveform.

More formally, the pulse plus tail is defined by its amplitude and phase content. The auto-correlation contains only the amplitude information, and the phase is lost. Thus, one can complete the estimation of pulse plus tail from the auto-correlation of the trace only by making an empirical assumption about the phase. It has been demonstrated that some parts of the chain of events are minimum phase, e. g., the peg-leg and reverberant multiples and some of the recording characteristics. Other elements such as the log of reflection coefficients itself, and pattern responses would appear not to be so. A minimum phase assumption is usually made.

If the model used in the design of the operator is close to the truth then deconvolution works well. If the tail is different from that estimated, or is absent, then it does not work. The output in the 'tail' region is no longer zero. A spike in the data, as from a static electrical disturbance, has no tail. The attempt at deconvolution simply results in the spike 'reading' the deconvolution operator and adding it to the recorded data.

Deconvolution usually adds more noise than it removes. Even in this case the output is generally an improvement for the interpreter. The noise removed is aligned noise, and the added noise is random. This comes about because, as described, the calculation of the input pulse and of the inverse operator to deal with it is not very accurate. Deconvolution operators for successive traces usually differ markedly, except at their leading ends. Their noise outputs differ from trace to trace, and it is generally nonaligned. Deconvolution achieves its effect first by reducing somewhat the amplitude of the pulse 'tail' (for which it was designed) and then obscuring the remaining low amplitude alignments by addition of random noise, and, most importantly, by giving reflections a more uniform character across the section and so improving their continuity. Deconvolution, in other words, reduces the variation in pulse form described in Level 2 chapter 8.

Thus it is quite possible for the deconvolution process to increase the noise level, and still improve the seismic section in the interpreter's assessment.

17.4 Normal move-out correction

The traditional treatment of NMO is based on a misconception which has yet to be revised in the modelling. This correction is one applied to reflection times to reduce the time observed at a distance X to the normal incidence reflection time which would have been observed from the same reflecting point.

Historically, the NMO correction, or spread correction, was calculated as a correction to a reflection time picked on the peak or the trough of a reflection pulse on a wiggle record so as to reduce it to the equivalent normal incidence time. A spike record is compressed at increasing source-detector distances, and it would be restored to the normal incidence version, the log of reflection coefficients, by the application of NMO corrections. If each spike is replaced by a seismic pulse, this relationship is lost. The pulses themselves are stretched by the application of NMO corrections, and so the corrected trace differs from the normal incidence trace, strongly at small reflection times, negligibly at large reflection times.

Where reflections pass into refractions at large source-detector distances and under some extreme structural conditions the seismic arrivals can get in the wrong order: in other words arrival time does not increase monotonically with depth, nor does NMO correction decrease monotonically with end trace reflection time. Where this situation is recognized that part of the data is discarded by muting.

Traditional NMO application stretches the data, usually so that reflection pulses become asymmetric, and leaves gaps in the data. These gaps are filled, either by some empirical rule, or by a more formal interpolation based on the assumption that the stretched pulse represents the signal. The NMO stretch represents a considerable addition to the aligned noise in the early part of the cross-section.

The traditional methods of applying NMO become particularly suspect when the interpreter compares the practical output of stacking with the ideal output to be expected from such a process. Thus for a 48-fold stack the ideal is

Increase of signal: random noise by 7 times.

Increase of signal: multiple noise by 48 times when the step-out difference between primary and multiple is about ¾ wavelength on the end trace.

This powerful increase of absolute values of signal over both kinds of noise would be remarkable: and the greater reduction of multiples should bury them still more completely in the remaining random noise.

No such performance is ever achieved, partly because only a small part of the noise is random, partly because there is insufficient step-out difference between primary and multiple and partly because of the inadequacy of the geological and seismic modelling on which 'stacking velocities' and step-out are determined.

The model is one with horizontal, isotropic layers. This is not very plausible geologically; and if it existed in nature it would attract no exploration interest. The model is restricted also to rather small emergent angles if even the model calculations are to have sufficient accuracy. In spite of this, the model leads to effective processing and interpretation when the departures from the horizontally layered earth are small – as in the large anticlines of Kuwait, Abu Dhabi and Saudi Arabia where the flank dips may be only 2° or 3°. It might be thought that more terms of an expansion, given by Koehler & Taner, are necessary to achieve a better fit for this horizontally layered model. Yet Al-Chalabi (1973) shows that when the ratio of spread length to depth is small the addition of a third term is all that is needed to obtain a high level of accuracy, and the addition of further terms does not necessarily reduce the error. When the ratio of spread length to depth is small, strong oscillations can occur in the residuals as the number of terms is increased.

18 T-samples and X-samples

18.1 Introduction: T and X

With the same rather critical attitude some still more fundamental aspects of current processing are examined from an interpretation view-point – still against the background that it works, and often well – but that the way forward seems limited by inadequacy of the modelling.

The question to be asked is why the seismic response of the earth is examined as time-samples at a given point, and not spatially. Already some radical departures have been made in the field of seismic holography: the material of this section is not so radical.

First, why was the choice made to use a time-series? There are two obvious answers. Digitizing the seismic trace from an analogue system yields a time-series. The processing of time-series is attractive, because, in the early days of development, it was able to draw on processing technology from other fields.

Yet the recording of digital seismic data is not a time-series, but an X-series along the spread at a given time (or nearly so), followed by another X-series one sample time later, and so on.

18.2 T-modelling

The signal content of a seismic trace (in the definition of 16.1), is a difficult series to refer to its real earth model. The data contained is the log of reflection coefficients. The line along which this log is taken is the locus of the foot of the normal to the reflectors, the bedding planes, throughout the geological sequence. In most geological environments this line will have the shape of a bent pin. At each angular unconformity there is a discontinuity in this locus: the pin is broken. In regions of fairly complex geology the locus will split into 3, 5 or any other odd number of branches. Continuing the analogy, the 'pin' will fork in such a territory. This last phenomenon is discussed more fully in a later section.

Such is the complex nature of a perfect spiked record, free from noise of any kind, and its model. In practice the recorded data is highly corrupt, from the several sources discussed elsewhere. Before the processor can begin to improve on the raw data, he must set up a model from which, by analysis, he can evaluate computational ways to reduce the noise. The model for a single trace is a model which involves the whole of the geological sequence, and the structure of every geological horizon. None of this is known: in fact, what is required to produce the necessary model is nothing less than the ultimate objective of a seismic survey, probably never to be attained.

The model which the data processor sets up is the familiar horizontally layered earth: usually with the further restriction of 'Goupillaud layering', in which each layer is of such a thickness that the two-way travel time normally through the layer is one time sample. Most seismic sections in areas of commercial interest depart very far from this model. In some cases the difference is so great that traditional processing can be regarded as providing little more than a first approximation to the model needed for control of the processing.

It is an awareness of these difficulties, even if not an explicit statement of them, which has led to other approaches to modelling and processing of seismic data. The frequency domain, the holograph and related concepts have all been used.

18.3 X-modelling

A seismic survey is conducted as a traverse along the surface of the earth – as is a surface geological survey. The geological survey finds, typically, a slow change of formation and of structure as it progresses. If the geological traverse is laid out along strike there is little change on the traverse except some facies changes.

The surface geological survey is equivalent to the seismic survey at zero time. The seismic survey can be viewed as a series of traverses, at successive reflection time intervals, over the same line of observation. In these traverses the rocks are known only by the seismic response, and not by visual observation. It is a purpose of this section and some later ones to examine these X-samples – the measures of particle velocity or pressure which are recorded at a particular time – arranged in order along the line of traverse. In particular, attention will be given to the geological modelling for this data, and some ways in which processes to improve the data can be derived from the X-model.

It is worth comment, in passing, that data are recorded on an X-traverse along the spread because of the multiplexing of the data from the several recording channels. It is a matter of some surprise that so little attention has been given to the processing of data in the very convenient form in which it is recorded.

For a deep, nearly horizontal reflection the model is very simple. The traverse may start late in the pulse on the near trace, and proceed to sample the same pulse at earlier points with increasing X. For shallower reflections, the more distant traces will work up the sequence a little into higher beds and their reflections. If the traverse is running down dip the traverse will work still higher into the sequence. A traverse up dip, where the dip is small, will first dip into deeper beds, then return to the near trace horizon and above it. Finally, a traverse up dip for a steep dip may extend into deeper beds throughout.

On each one of these traverses only a thin part of the stratigraphic column is involved, and the structure only of that thin part.

18.4 X-processing. Analysis programmes

The next step is to consider the processing which can be based on such a model. The first stage is to see whether analysis programmes can be used to define the model realistically before the processing of the data begins. Unless this can be achieved, even the small geological entity of the last section must be thrown back into the 'horizontally layered model' for want of knowledge.

The basic steps which the interpreter needs to have carried out are: –

I. Recognition of signal (cf Level 3, 16.1) and of aligned noise – reduction of the latter.

II. Determination of the X-shape of the reflection – a function of pulse form, velocity and dip.

II. will be examined first. If the X-data is re-arranged for a CDP gather – representing a traverse of source and detector both moving outwards at the same rate – it is seen that the dip effect on the X-shape is removed, and that the X-shape in the CDP gather is determined by the pulse form (or T-shape), and the characteristics of a thin part of the geological sequence. Variation in the pulse due to varying surface conditions and changing emergent angle adds noise to the X-shape.

So with a knowledge of pulse form and the stacking velocity one can envisage extracting signal from the CDP gather. If dip can be handled also then signal can be extracted from the recorded data direct, without re-arrangement as a CDP gather.

Implicit in this discussion is the use of velocity discrimination to accept or reject a seismic arrival as signal (cf Level 3, 16.1).

19 Some X-domain processes

19.1 Normal move-out application and stacking

One may observe the pulse form of a reflection, or one may assume that it approximates to a RICKER pulse or some other form. It is convenient to have this time pulse as some simple form. The stacking velocity, and hence the NMO is known. With this data, a template for a unit amplitude reflection can be prepared for each sample time of the normal incidence trace. The reflection has its peak at the sample time nominated where X=0, i. e. for the normal incidence trace, and the reflection time increases with increasing X in the usual way. This template consists of an X-series for that sample time, and other X-series at earlier and later sample times, so as to include the complete pulse form (typically 1½ cycles) throughout the spread. The X-trace is zero beyond 1½ cycles.

Application of NMO and stacking is achieved by applying this template to the data, cross-multiplying and adding to get an output. This achieves application of NMO and stacking in one pass. It is, effectively, determining an index of correlation between the template and the reflection.

Blanking at the smaller reflection times is achieved by zeroing the X-operators to correspond with the required blanking.

Figs. 52 and 53 illustrate the operator calculated for a particular configuration.

19.2 Pulse shaping and filtering

In earlier sections the case against long operators in the time-domain has been made for band pass filtering and for pulse shaping. These concepts are continued into the considerations of this section. The operator length in the time-domain will not exceed one and a half cycles of the recorded individual reflection pulse. The X-domain operators will be derived in the same way as the matched filter operators of 19.1.

The time-domain operators will be those required to convert the observed pulse to the required one in the T-domain. They are re-arranged as X-domain operators and padded out with zeros beyond the equivalent of 1½ cycles.

As before, this X-domain operator is applied to the data at the appropriate time, and the output is the sum of the cross-multiplications.

In a single pass one has now achieved NMO application, stacking and filtering, suitably blanked.

19.3 Introduction to a scheme for data processing in the X-domain

The foregoing sections are a digression from the work of the interpreter into some aspects of data processing. The excuse for this is that it serves as introduction to an alternative data processing routine. As always in this level one writes against the background that current processing works, while these ideas are untried.

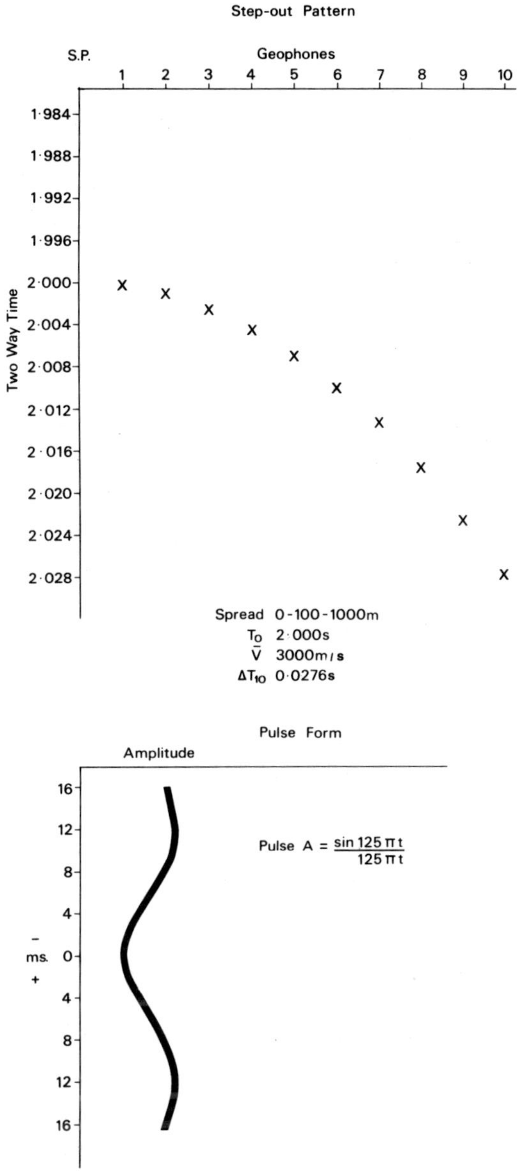

Fig. 52. Shows a step-out pattern and pulse form arbitrarily chosen for illustrative purposes.

One should summarize the reasons which might justify the pursuit of an alternative. First, time-domain processing in its traditional form seems to be nearing a limit to development, a fact attributed in earlier sections to inadequate modelling. Second, current methods involve a duplication of effort. The data processor must make decisions as to which reflections are primary and which are

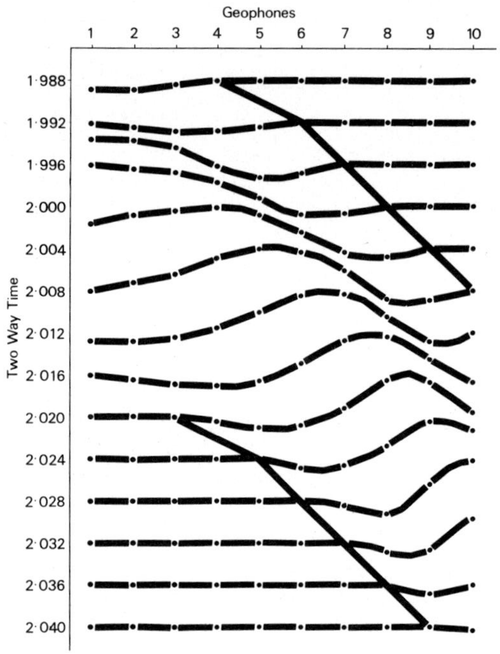

Fig. 53. The pulse and step-out of fig. 52 plotted as amplitude against distance in the CDP gather.
This matrix serves as a two-dimensional operator. It is applied to the data of the CDP gather, and corresponding elements are cross-multiplied and summed to give a correlation index. This index, suitably scaled, is the output for the operator time – in this example, at 2 s. Outside the limiting lines shown the values in the matrix are zero.
This process has applied NMO, stacked the data and applied a matched filter in the time and K-domains. By suitably modifying the operators, any other convolution type process, either in time or K-domain, can be accomplished in the same pass.

multiple. He must decide whether certain values of interval velocity are geologically plausible or not. His decisions are usually right, but are communicated (if at all) in computational vocabulary which is not readily taken as a starting point for the interpretation: so the interpreter makes a fresh start. Finally, the interpreter will sometimes doubt whether the decisions taken by the processor are correct ones, and he may call for re-processing. And always he will be pressed for time, wishing to see the data in a useful form as soon as possible, and before any impact of processing decisions has been seen on the data.

The approach to processing outlined in this chapter is an attempt to meet these demands.

19.4 The velocity survey

A first requirement is a form of presentation which can be made without any preliminary processing decisions. What is proposed is a two-fold display.

The first is a display of power traces for each CDP gather. These are the outputs from velocity determinations made by stacking a CDP gather with a wide range of stacking velocities. The power in a series of time windows is determined for each velocity. Two plots are normally derived.

I. A plot of power against stacking velocity for a constant time.
II. The maximum of each trace of I. plotted against the time.

This last is the so-called power trace.

The display envisaged is the power traces for every gather arranged in order. Every maximum of each power trace is annotated with the ΔT value which yielded it, and the stacking velocity corresponding to it. Corresponding to each power trace maximum is a window of a stacked trace which yielded that maximum power value.

The second display takes all of these stacked windows corresponding to the power maxima of the power trace. The windows may overlap, in which case the overlap region is merged. Or there may be a gap between windows, in which case the trace is zeroed.

The traces of this display will thus show all of the principal reflections stacked with the optimum step-out values. The traces will also show all of the principal noise (mostly multiple) alignments, stacked with the step-out values optimum for them.

19.5 Principal use of these displays

Because these displays require no preliminary analysis programmes, and no processing decisions, they can be run as a matter of routine, day by day, as soon as a crew returns to its base after the day's work in the field. They will be available for inspection late in the evening or the following morning. They will give added confirmation that equipment is functioning, and will help to demonstrate that the exploration objectives are being met.

This description envisages that the crew is equipped with an adequate computer. The further the processing facilities are removed from the crew's base, the longer the delay which is interposed.

These displays will be adequate for a reconnaissance interpretation. The power trace display permits the tracing of rather gross continuity, and the annotation of step-out and velocity offers other evidence for the continuous character of the event.

The 'window' display offers more precise continuity and reflection timing. With these more precise times and the step-out values, interval velocities can be calculated, and the reflection sequence related to the geological sequence – known or inferred. The examination of velocities supports the interpreter's discrimination between primary and multiple reflections.

The interpreter will finish with these two displays marked up with primary reflection continuity, leaving all other events unmarked. For a number of reconnaissance purposes these displays will provide all that is needed for structure or stratigraphic studies. It will, however, generally be necessary to progress to a final section. In the further discussion it will be seen that the preliminary interpretation has made most of the processing decisions.

The decision which remains to be taken is that of the final pulse form. This is also a decision best taken by the interpreter to meet his needs. If the interest is solely in good continutiy, he will specify a pulse form (varying with time) which matches the reflections that he wishes to follow. If the interest is in resolution for all or part of the section, he will specify a narrower pulse. As a general guide it can be said that if the zero-crossings of a reflection are smoothly aligned, the centre frequency of the pulse can be doubled by a deconvolution spiking operator without increasing the noise level too far to obscure the reflection. On the other hand, if the zero-crossing alignment is ragged, the centre frequency cannot be increased so far.

19.6 The proposed scheme of processing: preliminaries

The interpreter's picks on the second display show all the primary reflections of significance. These picks can be digitized on a digitizing table, and this record serves as a starting point for many investigations. Perhaps the most valuable is that one can then use these digitized picks to re-enter the data in the form in which it was plotted, and locate the reflection time on peak or trough, either to the nearest sample, or by interpolation to greater accuracy. By reference to the first display, the appropriate step-out and velocity values can be determined for each primary reflection time.

Of great value, also, is the observation that any power trace maximum not accepted by the interpreter is of necessity aligned noise.

Since the aligned noise obscures the signal for the interpreter more than anything else, first attention is given to its reduction. The first displays identify this aligned noise as those power maxima not recognized as primary reflection signal by the interpreter and shown on the section by his picking of continuity, and they tabulate its step-out. The waveform of the noise is seen in the 'window' display, where it is stacked with this optimum step-out. A number of methods are available for reduction of aligned noise when this much is known about it. An example is the method "Souston" of MICHON, WLODARCZAK & MERLAND (1971).

The noise waveform seen in the stacked window is now to be subtracted from each of the CDP traces from which it is derived. Two steps are necessary. First, the waveform is shifted to the appropriate time position on the CDP trace – a position determined by the NMO value on the power trace display. Should the CDP traces of the gather vary significantly in amplitude a scaling factor for each trace is used to match maximum amplitudes in the waveform to be subtracted and the waveform in the recorded data of the CDP gather.

The next stage is to establish the stacking velocity-time relationship for each gather. The basis for this is that each primary picked by the interpreter yields a reflection time T and a stacking velocity $\overline{V}$. Between observations at different times for the same gather the assumption of constant interval velocity gives a straight line in the plot of $\overline{V}^2T$ against T. Extrapolation upwards from the shallowest reflection gives a straight line through the origin – a constant value of $\overline{V}$ which is that determined for the shallowest reflection.

Extrapolation downwards of the $\overline{V}$ curve below the deepest observation cannot be based on observation, since there is none. Two alternatives are possible.

a) If there are reasonable grounds for postulating the geological sequence below the deepest reflection down to the deepest level of interest (usually the bottom of the section), and if the interval velocities can be postulated with reasonable accuracy, then these values can be used to extend the $\overline{V}$ curve.

b) If there is no such information available, then the extension can be made only on a statistical basis. It is shown in 19.9 that this leads to an extrapolation of the form;

$$\overline{V}=b\,T^{1/5}$$

where the constant b is determined for the $\overline{V}$ and T values of the deepest reflection.

It is this completed $\overline{V}$ curve against T which governs the step-out pattern of the X-operators or reflection templates to be used.

The steps taken before the processing run can now be summarized.

I. Data gathering and documentation.

II. Decision on blanking pattern from inspection of monitors.

III. The 'reconnaissance run' of para 19.4.

IV. Interpretation of these displays, tracing all primary reflections.

V. Decision on pulse form required in output.

VI. Processing to attenuate aligned noise, as in 19.6.
VII. Digitizing of the picks under IV. above.
VIII. Construction of $\bar{V}$, T tabulation for each gather, including its extrapolation.
IX. Construction of X-operators (as described in 19.1 and 19.2), for each gather, at each sample time.

19.7 Smoothing of velocity values

It is a common practice in traditional processing to smooth stacking velocity values. There is full justification for this. The velocities determined are subject to experimental error. The 'geological space' in which the seismic wavefronts or rays travel for successive CDP gathers shows a strong overlap, and physically, this must smooth any changes due to lateral variation in the rocks. Finally, these changes will themselves generally be small since facies changes are often gradual and usually restricted to a small part of the geological sequence.

The smoothing is often applied to the stacking velocities calculated for each sample time along the traverse. Sometimes the interval velocities are calculated, these values are smoothed, and the stacking velocities $\bar{V}$ are re-calculated from them.

It seems more logical to smooth the step-out values observed on a continuous reflection. This is an observed quantity, and from geological and geometrical considerations it is clear that it must change smoothly along the traverse.

Smoothing a continuous series of step-out (ΔT) values can be applied equally well to reflections which are many-branched in geologically complex areas, and produce forked $\bar{V}$, T curves. In such cases the $\bar{V}$ may have several values at a time T, and smoothing of $\bar{V}$ is no longer meaningful. The concept of interval velocity, also, loses significance when the reflection times and step-out values to a single reflector can have several values at an observation point.

19.8 The processing run

After the extensive preliminaries of 19.7, the processing run is simple to describe. It consists in applying the templates, or X-operator group, to the data. The correlation index: that is, the sum of the cross-multiplication products, is the output of the sample time – one value for the stacked trace.

19.9 Other uses of the 'reconnaissance' displays

There are other uses for the digitized record of the interpreter's picks.

These digitized picks – if proper identifications are added – can be combined with topographic survey data on the shot points and geophone positions. From this combination can be generated a plot tape to plot the reflection times on maps of selected horizons. Alternatively, contours can be generated and plotted.

The same digitized picks can be used as input to further analysis programmes. Detailed interval velocity studies can be carried out. A further reference back to the data will call out the amplitudes of the reflections also: and amplitude-velocity changes bear on the estimation of density and fluid content.

Still another use is to contribute the digitized data to a data store. This can be used to generate regional maps, or to up-date existing maps if there is already seismic coverage in the area.

19.10 Migration

An interesting possibility exists for the direct migration of field data, using the concepts of these X-operators.

Consider first a field record: with static corrections applied if necessary, having a reflector of uniform dip. The hyperbolic shape of this reflection will be the same as if the reflector were horizontal, but it will be displaced up dip. The static corrections can be applied in a time-domain process, or more simply by reading the X-data along a time-line which is not a constant time, but a time-line which expresses the static corrections. The reflection seen at some point on the flank of the hyperbola on the inner trace of the field record should be at the vertex of the hyperbola in its migrated position.

Consider then the group of X-operators which is used to derive a correlation index from a gather, and so output one sample value of the stacked and processed trace. Extend these operators in both directions: the distance will be calculated from the dip which is set as the limit of interest: the sample interval will be that of the trace spacing.

In simple terms, the exercise is one of sliding this group of X-operators, used as a template, across the data from side to side until the template is in register with the recorded reflection. This identifies the amplitude of the reflection and the time and distance values to which it is to be migrated – its position on the migrated section. At the same time one would like to carry out the other functions of these operators as described in earlier sections.

This is achieved by summing the cross-correlation of each extended X-operator with its corresponding recorded X-series. The maximum value of this summed cross-correlation is the required amplitude (usually all amplitudes so derived must be scaled down). The time is that for which the X-operators were calculated. The displacement is the number of traces displacement required to achieve the maximum of the cross-correlation output.

There are two sources of error in such a procedure. Section 17.3 deals with the difference between migration and stacking 'velocities'. The effect of curvature of the reflector is to change the curvature of the hyperbola recorded on the field record, while there is no corresponding change of curvature on the CDP gather. This again will lead to differences between stacking and migration velocities – between the recorded reflection shape and the template. In addition, migration as a data process suffers from the disadvantages already listed in Level 2, 9.6, and elsewhere.

19.11 Extrapolation of $\overline{V}$ in depth

19.11.1 The problem

Logically, there is no way to infer the velocities in depth from a set of velocities measured at shallower depths. Expressed in geological terms, one cannot derive the properties of the Mesozoic rocks by making measurements in the overlying Tertiary.

The problem remains, however. The data processor needs to apply some values of ΔT at reflection times greater than those for which he has observations. The interpreter, similarly, may need to derive the structure of reflections below the established velocity control levels; in other words, below the levels at which reliable velocity determinations can be made, even if reflection alignments can be detected.

19.11.2 Some evidence

Two approaches are possible. If geological information, or good inferences, are available; and if these can be expressed reliably in terms of seismic velocity, then this data can be used to extend velocity and ΔT determinations in depth, at least over part of the range of extrapolation interest.

In the complete absence of factual guidance, the data processor requires to use ΔT values which have the highest probability of revealing deep reflections, should any exist in the recorded data.

To meet this requirement one must turn to the statistics of the relation between velocity in rocks and their depth of burial.

Faust (1951), on the basis of extensive statistical studies in North America derived the relationship that

$$v_z \; \alpha \; Z^{1/6}$$

and it is remarkable that Gassmann (1951), working theoretically on velocity through packed spheres also arrived at the expression

$$v_z \; \alpha \; Z^{1/6}$$

In both cases the expressions are modified somewhat by the properties of the fluids in the pore spaces: but for the present purpose these minor modifications are ignored.

The proposal is, therefore, that in the absence of any geological guidance, the calculation of velocities and ΔT values below the levels of observation, be based on the relationship

$$v_z = a \, Z^{1/6}$$

where a is a constant.

This leads readily to the relation

$$\overline{V} = b \, T^{1/5}$$

as shown in (3) of section 19.11.3 where $\overline{V}$ is the time-weighted r. m. s. velocity – the stacking velocity, T is reflection time, and b is a constant$=0.70940 \, a^{6/5}$.

Thus, from the deepest reliable determination of stacking velocity at time T one can determine the constant b, and use this to extend the $\overline{V}$ curve or table in depth, and so derive the most probable values of velocity and ΔT.

From the value of b one can also derive the value of a: and from the relation

$$v_z = 0.83938 \, a^{6/5} \, T^{1/5}$$

one can check the implied interval velocities.

19.11.3 Formulas: The FAUST-GASSMANN relationship

Summary of formulas

(1) $v_z = a\, Z^{1/6}$ fundamental

At time T

(2) $v_z = 5^{1/5} \cdot 12^{-1/5} \cdot a^{6/5} \cdot T^{1/5}$
$= 0.83938\, a^{6/5} \cdot T^{1/5}$

(3) $\overline{V} = 5^{7/10} \cdot 7^{-1/2} \cdot 12^{-1/5} \cdot a^{6/5} \cdot T^{1/5}$
$= 0.70940\, a^{6/5} \cdot T^{1/5}$

(4) $V_a = 5^{6/5} \cdot 6^{-1} \cdot 12^{-1/5} \cdot a^{6/5} \cdot T^{1/5}$
$= 0.69948\, a^{6/5} \cdot T^{1/5}$

(5) $T = 12 \cdot 5^{-1} \cdot a^{-1} \cdot Z^{5/6}$
$= 2.4 \cdot a^{-1} \cdot Z^{5/6}$

(6) $Z = 5^{6/5} \cdot 12^{-6/5} \cdot a^{6/5} \cdot T^{6/5}$
$= 0.34974\, a^{6/5} \cdot T^{6/5}$

20 Some possible refinements of X-operator processing

20.1 Introduction

X-domain processing appears to offer some opportunities for refinement which are readily incorporated. A general theme is that as more is learned from the results of analysis programmes about the configuration of reflections in general, or about a particular reflection of interest the modelling can be improved, and the template or X-operators to be used can be similarly improved.

Two such possibilities for refinement are discussed in the following sections: they seek to reduce the error due to multiplexing the field data, and the error due to the assumption of a hyperbolic form for a reflection T-X curve in place of an observed relationship.

20.2 Multiplexing error

The recording problem for exploration seismics is that a number of channels are generating waveforms, usually 24 or 48. These cannot be sampled simultaneously at a given sample time t without providing digitizing and recording equipment for each channel: an impractical idea. The established technique is to multiplex the data: to sample and digitize the channels in turn: starting at sample time t on the first trace, then at successive short intervals at the 2nd, 3rd . . . 24th or 48th traces. These short intervals are such that at time t+1 (where time is measured in these short intervals) the equipment takes the next value from trace 2, and so on. It will be apparent that if the sample time interval is 4 ms, the end trace will be mis-timed by almost 4 ms (in practice 3 ms in most equipment), and will be recorded as earlier than it is.

The corrective here is applied in the conversion of time-domain pulse forms or operators to their X-domain equivalents. If multiplexing error is ignored, the waveform to be converted is sampled at

$\Delta t_1, \Delta t_2 \ldots$ for 1st template
$\Delta t_1+1, \Delta t_2+1 \ldots$ for 2nd template
.
.
.
$\Delta t_1+r, \Delta t_2+r \ldots$ etc for $(r-1)^{th}$ template

where Δt_1 Δt_1 . . . are the successive NMO corrections.

r has a range of integral values, positive and negative, to complete the template for the reflection pulse across the gather, the negative values of r being required if time Δt_1 on the first template is after the leading edge of the pulse.

Time here is measured in sample intervals.

To take account of multiplexing error, the waveforms are sampled at

$\Delta t_1, \Delta t_2+\delta t, \Delta t_3+2\delta t \ldots \Delta t_p+(p-1)\,\delta r$
$\Delta t_1+1, \Delta t_2+1+\delta t \ldots$
.
.
.
Δt_1+r, $\Delta t_2+r+\delta t$ . . . and so on, by an interpolation procedure if one is considering sampled waveforms, by calculation if one is considering a continuous waveform or template as in 19.1

20.3 The T-X shape of a reflection

The earliest assumption as to the shape of this curve was that it was a parabola, a crude assumption adequate only for short spreads. A first stage of refinement was the hyperbola, the assumption of a horizontally layered earth, and a restriction on the angle of emergence.

Real earth observations do not fit either assumption very well, even when the beds are almost flat-lying. Most observed forms appear to lie between these two shapes, and a little closer to the hyperbola. The question is an important one, for it can be shown that for a typical field configuration, a ΔT error corresponding to ¾ of a wavelength reduces that frequency component of a reflection in the stacked trace so much that the signal to random noise ratio of the stacked trace is that of a single trace of the gather. In other words, the gain over both random and aligned noise is gone.

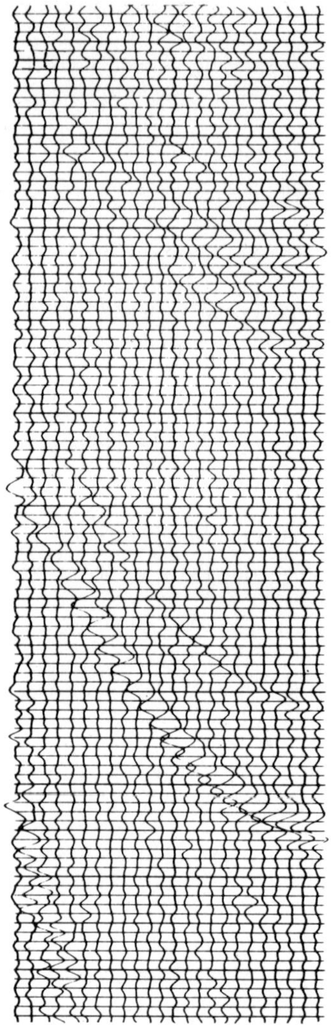

Fig. 54. The result of summing 12 marine records, after correcting arrival times for small differences due to statics and to structural changes. The T-X curve is neither hyperbola nor parabola, but some intermediate shape.
Note also the change of pulse form with increasing value of X. The pulse becomes wider with increasing X. Application of NMO still further stretches the distant traces, so that the mismatch when they are aligned for stacking is made worse.

Experimentally, the true shape of a reflection in a T-X plot can be examined from the records of a good reflection in a flat structural environment. It is necessary to reduce the random noise to the point where it is not significant, where the resulting errors of timing are, say, less than 1 ms. The procedure is to take 20 or 30 consecutive records of this kind: apply static corrections, if any: and a 'structural correction'which is to bring all of the traces used to the same structural level. These 'structural corrections' are determined by taking smoothed reflection times from a stacked section, and using as correction the difference between these reflection times and that of the first trace used. All of these corrected field records now show that shape of a reflection from a horizontal bed, with some second order errors. These are the residual static and structural errors, and changing step-out values along the line of traverse resulting from change in velocity.

Addition of all first traces, and so on, produces a record in which time differences are those of step-out, with only negligible errors. Fig. 54 is an example of such a record. It has a 7 or 8 ms departure from either the parabola or the hyperbola of best fit. This measure of departure does not leave a large margin for error in stacking velocity before stacking advantages are lost.

20.4 A process for determining the T-X curve with some precision

A method is described for determining the shape of the T-X curve, based on the experimental method of 20.3. It is illustrated first for reflection time data, and then for waveforms. It is essentially an approach used by MORGAN (1970) and others, though with a different objective.

The reflection times of multiple cover data for a single reflection are arranged in a matrix with shot-point numbers running horizontally, and geophone numbers vertically. This is illustrated in table 1.

Table 1

		Shot-point					
		11	12	13	14	15	16
Geophone	1	1.996	2.008	2.016	2.018	2.017	2.022
	2	2.003	2.011	2.018	2.022	2.026	2.027
	3	2.009	2.019	2.022	2.027	2.031	2.036
	4	2.020	2.030	2.033	2.032	2.038	2.047
	5	2.030	2.039	2.044	2.041	2.044	2.052
	6	2.038	2.048	2.054	2.055	2.059	2.061

Common shot-point: vertical line	11/1	11/2	...	11/6
where 11/1 is S. P. 11, geophone 1				
Common offset: horizontal line	11/1	12/1	...	16/1
Common geophone: oblique line	11/1	12/2	...	16/6
Common depth point: oblique line	13/1	12/3	11/5 – A three-fold stack	
or	13/2	12/4	11/6	

Analysis of this block of data by the method described yields the following results (using some other trivial manipulations to estimate values where the oblique line of data is incomplete).

S. P.	Comparative reflection time (s)	Comparative S. P. static correction (ms)	Comparative geophone static correction (ms)
5			−3
6			+1
7			+5
8			+4
9			0
10			−1
11	1.999	+4	−1
11½	2.005		
12	2.013	+5	0
12½	2.013		
13	2.022	+5	+1
13½	2.022		
14	2.026	+3	−2
14½	2.027		
15	2.029	+3	+1
15½	2.030		
16	2.031	+3	
16½	2.033		
17	2.035		
17½	2.040		
18	2.043		
18½	2.044		

Geophone position	NMO (ms)
1	0
2	4
3	6
4	10
5	17
6	26

This data can be sliced four ways. A horizontal alignment represents a common geophone number – common offset position, and therefore a common ΔT value subject only to change with the value of T_0 and $\bar{V}$. A vertical alignment represents a common shot-point, and a common shot-point static correction. An oblique alignment can be found which represents the common depth-point data, a common normal reflection time. Finally, an oblique alignment can be found with opposite slope which represents a common geophone position, and a common geophone static correction.

One can proceed with the following steps

I. Take the mean reflection time value along the third direction – the common depth-point. This gives a mean normal reflection time, plus the means of static and dynamic corrections. The static corrections are taken as random: the dynamic corrections as almost constant from one third direction out to the next. The differences of successive mean values along the traverse represent changes in normal reflection time, i. e. structure.

II. From these changes 'structural corrections' can be calculated and removed from the data. The matrix is now one of static and dynamic corrections, plus a constant reflection time.

III. Differential shot-point and geophone static corrections are determined in a similar way by slicing in the first and fourth directions, and these differentials are subtracted.

IV. Finally, horizontal averaging, over 20 or 30 traces at a time where the beds are fairly flat, will yield the required T-X relationship.

If only the T-X relationship is of interest, this can be determined directly as a first step in the sequence.

Reflection times cannot be used if the data is not of sufficiently good quality for single cover reflections to be identified and timed. But if they are seen on a stacked section, then time windows from the corrected gather can be assembled and placed in the same arrangement as the reflection times. The peak of the stacked reflection is chosen as a local zero, and this same zero is used for all of the windows from the corrected gathers.

The routine is as before: except that instead of taking a mean, one sums the traces. The output shows a peak which may not match the common zero. This indicates a revision to structure, static or dynamic correction, to be applied to those values which yielded the stacked section which was input to the process.

Essentially this manipulation is used in a recent paper by TANER et al. (1974).

20.5 Use of T-X curves

However T-X curves are established, they give data for revising the shape of the templates or X-operators. It is probable that the T-X curve is best expressed as

$$T_x^2 = T_0^2 + \frac{X^2}{\bar{V}^2} + AX^4 + BX^6 \ldots$$

and that A and B will show a measure of time and space variance.

21 Interpretation and the frequency domain

21.1 Frequency domain interpretation

Geologist and seismic interpreter are familiar with a time-domain presentation of data, in which distance and time are similar enough to distance and depth to make visualization relatively easy. A presentation of frequency and phase has no such visual similarity to the real earth; but the data is so often cast into this form that it may be of interest to examine the little that the interpreter can extract. It is not much.

There are a few models for which the frequency domain response can easily be visualized.

A horizontally layered earth (so useful to the geophysicist) gives identical amplitude and phase spectra for every trace with the same offset distance. There are several ways of presenting frequency domain data along a line of traverse; one way is to present the amplitude and phase spectra, plotted against frequency, for successive traces. Alternatively, each FOURIER component can be represented as a vector, its length being given by the amplitude a, and the orientation by the phase φ. In the horizontally layered earth each frequency vector would remain unchanged along the line of traverse. Any other vector presentation can be used, such as separate plots of $a \cdot \sin \varphi$ and $a \cdot \cos \varphi$.

Inspection of these vectors for a real traverse along flat-lying rocks will show how inadequate the model is for frequency domain work.

A dipping, but regularly layered earth will show the vectors staying at constant length, but rotating along the line of traverse at a rate proportional to the frequency. See fig. 55. Some additional slow changes are introduced, however, by the incoming of new beds at surface down dip, and by the exclusion of some older beds from the sample data down dip.

A perfect angular unconformity produces two sets of vectors, one above and one below the unconformity. Each retains constancy of vector length, but they rotate at different rates along the line of traverse and their sum is complicated. Even with this simple model the vector behaviour is hard to visualize in terms of structure: and any further approach to real earth models produces vector behaviour which it is impossible to visualize in structural terms.

21.2 A noise criterion

At the surface spacing of seismic observations, most geological changes take place slowly. In frequency domain terms, the rate of change of the vector will be slow. Any rapid changes can be classed as noise. The rate of change which must be taken as limiting is defined by the dip.

This concept is one which is only of academic interest at the moment. It may justify re-examination when many more channels of information are recorded.

It can be shown that most FOURIER components of a seismic section do rotate and change length in a regular fashion along the line of traverse, overlain by much irregularity. If the vector behaviour is smoothed (by smoothing $a \cdot \cos \varphi$ and $a \cdot \sin \varphi$), and then if the smoothed vectors are recombined, very good seismic traces, apparently with low noise, result. A little work has been done in this direction, using common offset traces to preserve the same geometry.

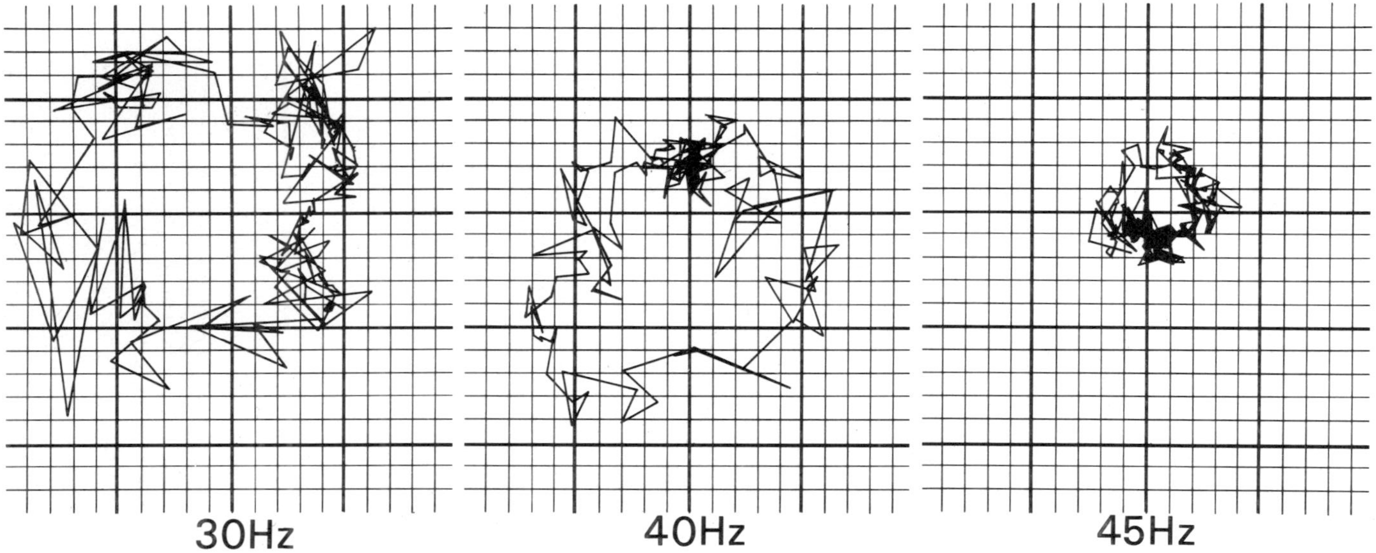

Fig. 55. Vector plots of FOURIER components of the far trace along the line of traverse. 30, 40 and 45 Hz components are shown.
The sequence is gently dipping. The FOURIER component vectors are therefore expected to describe circles. It is concluded that the circles here contain the geologically significant information, and that the superimposed irregularities are noise.
Considerations of this observation lead one to consider smoothing these FOURIER component vector plots, and then recombining the components to give low-noise traces in the time-domain. FOURIER components of high frequencies are sometimes so noisy that they do not lend themselves to smoothing procedures, and perhaps they should be discarded in any case. High frequencies, or high dips, can give rise to alias problems in this treatment.
The fundamental observation is that geological changes along the line of traverse are mostly gradual ones. Even where the geological change is abrupt, as at faulting, the seismic change is not: the wave front continues to advance. Abrupt seismic change along the line of traverse is noise: this may point to some alternative methods of treatment.

22 Amplitude-distance relationship, and pulse form

22.1 The model

The model is a plane interface between two formations, with some stratigraphic detail showing seismic contrasts, above and below. The amplitude of this reflection is measured along a regular spread.

Three factors govern the change of amplitude

I. The path to the further geophones is longer than the path to the near ones, except in the case of shooting up dip. In the latter case, the path length will decrease initially if the dip exceeds a certain small value: the amplitude will reach a maximum, and then decrease for a considerable range of dip above this value. For large dips the amplitude will increase throughout the spread. The distances at which these changes occur are dependent on depth and dip of the reflectors and the spread configuration. This is the phenomenon of spherical spreading: amplitude is roughly inversely proportional to path length.

II. Reflection coefficient increases with the angle of incidence. Normal incidence reflection coefficients are given by the formula of 3.2 Level 1. As the angle of incidence increases, the reflection coefficient may first decrease, it then increases at first slowly, and then much more rapidly as the critical angle is approached: at the critical angle (for P-waves) the theoretical reflection coefficient is 1.

III. It has been pointed out in 3.6 Level 1 that the reflected pulse form is to be regarded as the convolution of the incident pulse with the 'microstratigraphy', expressed as a log of reflection coefficients, to give the reflected pulse form.

As the angle of incidence increases, the interval (in time) of these minor interfaces increases: and the individual reflection coefficients increase or decrease. Both change in a functional but complicated fashion. Thus the comb function with which the incident pulse is convolved changes with the incident angle. This impresses a change of amplitude with incident angle, and so with distance. It impresses also a change of pulse form with distance, which may also prove to be of value.

The effect of these three factors is to produce a change of amplitude with distance, and this relationship is a characteristic of the reflection.

22.2 Use of a-X curves

The relationship of amplitude with distance for a single field record is usually too irregular to be used. It is, however, possible to take up to 20 consecutive records in an area of flat-lying sediments, and to sum the a-X curves, thus increasing the signal to random noise ratio. In some cases, the a-X curve has a characteristic form, which is repeated with little change with the next 20 fields records. The relationship is thus, potentially, a basis for correlation of reflections across an interruption in continuity.

The relationship is one, moreover, which will afford some evidence of facies or fluid content changes within that small part of the sequence which contributes to the seismic reflection pulse being examined. For example, the change from a thin water sand (< 30 m) to a thin gas sand would be expected to make a marked change in the a-X curves, as well as a marked change in the absolute amplitude of the reflection derived from the complex.

It is interesting to recall an early paper by KOEFOED (1960) who called attention to this relationship.

22.3 The problems of studying pulse form

Amplitude changes are accompanied by pulse form changes, which have been studied but little, although the changing pulse form has been studied by MORGAN (1970) in a configuration such as that of 20.4.

The problems of pulse study are twofold. First is that of getting a form for study which is only slightly corrupted by noise (in the definition of 16.1). Second is that of achieving resolution of the information content of the pulse. Clearly the second can be attempted only when the first has already been achieved.

22.4 The observation of pulse form

What is the form of the seismic pulse? This question has exercised geophysicists for some years, as the references in the bibliography will show.

The pursuit of the seismic pulse is based on the view that the seismic record is the convolution of the log of reflection coefficients with a pulse. If this pulse were known, its inverse would be known, and so the log of reflection coefficients could be recovered. It has already been seen that this is an over-simplification. There are so many factors (as described in 3.5 of Level 1) that influence the form of the seismic trace that any attempt to explain the seismic trace in terms of two factors only will fail to give a precise solution.

Nevertheless, the problem remains. Progress towards resolution of the information in the seismic pulse observed depends on the approximation which can be made of the pulse form.

Perhaps some progress can be made on these lines. One estimate of the 'seismic pulse' which has travelled through a nearly uniform rock is the 'RICKER pulse' observed in the Pierre shale. This is a functional form, expressed as

$$\psi = \left(1 - \frac{T^2}{2}\right) e^{-\frac{T^2}{4}} \quad \text{(MUIR \& HALES, 1955)}$$

and it has a smooth amplitude and phase spectrum.

A much closer approach can be made, based on the experimental results of O'BRIEN (1969). This paper presents the observed shapes of seismic pulses at short distances (mostly between 20 and 80 m) for charges of various sizes in clay and sandstone.

A review of the recording and processing procedures which affect pulse form shows that they also are functional and have generally smooth spectra.

The real earth contributions are three. First there is transmission through a layered earth which involves many peg-leg multiple reflections. The earth does not have strong resonance in the geological sequence, though there may be local periodicities, as in a grey-wacke sequence or in the cyclothems of a coal-measure sequence. A typical reflection coefficient log approximates closely to a sample of a random series, with its spectrum having a ± 3 dB variability if there is no smoothing. It is a generalization that the transmission operators will tend to have reasonably smooth spectra, but with some irregularities due to local periodicities in the sequence.

Second is the effect of the near-surface geology. The thickness of material involved is small, and it varies in thickness and in detail from one observation point to the next. The consequent near-surface 'operators' are generally spiky, and rapidly changing along the line of traverse. The spikiness is reduced a little by the fact that there are two near-surface operators, at shot-point and receiving array, and each of these may be made up of multiple sources and detectors: variations will occur among all of these, so that the cascaded operators will be smoothed a little. In general however they will remain somewhat spiky, at least for most land surveys.

Finally, the reflector operator, the expression of the microstratigraphy, is concerned with a small geological sequence only, and its spectrum will be very spiky also. Changing the angle of incidence will change the form of the operator and its spectrum, as described in 22.1, so that stacking reflections of different angles of incidence – different offset distances – will reduce the spikiness.

These considerations lead to a view that the spectrum of a seismic pulse, or of a sequence of seismic pulses, can be regarded as having two components, one smooth and one spiky. The smooth part carries mostly information about the ideal pulse, recording and processing parameters, and seismic transmission. The spiky part carries information mostly about the near-surface and the log of reflection coefficients.

Smoothness and spikiness need clearer definition. One can fit an analytical form, such as a Ricker pulse spectrum, to the observed spectrum as the smooth part; and assign the departures as spikiness. Smoothing operators can be applied to obtain the smoothed spectrum. Or one can take the spectrum of the spectrum, in which the smooth part appears as the low frequencies and the spiky part as the high frequencies, often with a pronounced minimum between them, which is a convenient separation. This last approach leads deep into spectrum analysis and some phase problems.

However, even when the spiky part of the spectrum is isolated, its information content in land surveys is nearly as much of surface origin as it is of the subsurface. It will also contain noise and residuals from other operations (in so far as they do not have smooth spectra). The remaining problem is that of increasing the ratio of subsurface information to surface information.

The proposed route is

I. Use common offset traces, to avoid the losses on stacking described above.

II. Remove structural effects by applying corrections to remove time differences arising from this effect.

III. Sum groups of traces. Reflections are in phase, so that signal to random noise will be increased.

IV. Surface effects will be changing along the line of traverse, so their contribution is reduced.

Ideally this should be tried with parts of traces bounded above and below by continuous reflections. The outputs from III., with only the spiky parts of their spectra presented, would appear to offer some chance of approaching the log of reflection coefficients. Any other method of spiking the data effectively would also do this.

A repetition of the procedure with other offset traces allows the application of criteria developed in para 22.1.

If it works, the approach of this section would seem to lead to a refinement of the stratigraphic interpretation of seismic data in terms of facies change and fluid content change, particularly over thinner stratigraphic units.

22.5 Resolution

If any of the many transfer functions in the long chain that leads to the seismic section have a high cut characteristic, then the high frequencies will not be recorded, and there is no way to recover them. The three transfer functions here considered are those of earth transmission, sampling, and anti-alias filter.

Most geophysicists take a pessimistic view of the transmission characteristics of the earth, and one is familiar with seismic sections limited to 30 or 40 Hz at the upper end. Yet one must recall the penetration achieved at sea with kHz signals; and some French work in collieries with these high frequencies. If the appearance of a seismic trace shows it to be dominated by 40 Hz signal, this is no evidence of the signal to noise ratio at 400 Hz. Even if it were 1 : 1 at 400 Hz, the component would probably be lost in recording and processing procedures which are not precise enough.

Sampling rate sets an absolute limit to the pass-band. 4 ms data is limited at 125 Hz and is folded back above that frequency. Any frequencies above 125 Hz which are present will thus distort the data unless they are removed before sampling. Clearly, where resolution is of interest the sampling rate must be increased. The highest rate commonly in use is 1000 samples per second.

Anti-alias filters are designed to have a high attenuation at the folding frequency: an often used specification is 80 dB down at 125 Hz for 4 ms sample interval. Such filters may take an octave to achieve the required attenuation, so that there is a progressive cut-off from 62½ to 125 Hz. Only negligible amplitudes are folded back from 125 Hz.

It has often seemed that the extreme caution shown in the design of alias filters is not justified. For example, if the anti-alias filter has a cut-off at 83⅓ Hz, and achieves 80 dB attenuation at 166⅔ Hz it is seen that with a folding frequency at 125 Hz all data above 83⅓ Hz is corrupt because of the folding, but that there is no attenuation up to that frequency. This would seem to offer a little wider frequency range in the search for resolution: and its impact at higher sampling rates is stronger.

22.6 The one-dimensional synthetic seismogram

An interesting application of knowledge of the form of the seismic pulse is in the preparation of synthetic seismograms. I am indebted to Dr. P. N. S. O'Brien for the text of this section.

If the density of the subsurface, ρ, and the velocity of sound through it, V, is known continuously as a function of depth, then it is possible to calculate the exact form of the ideal, noise-free, seismic reflection trace (PETERSON et al. 1955; BARANOV & KUNETZ 1960). Such a calculated trace is called a Synthetic Seismogram. The calculation proceeds by approximating the subsurface, with its continuously varying density and sound velocity, by a series of layers in each of which the density and velocity is constant. The reflection coefficient at the interface between successive layers is given by

$$(\rho_n V_n - \rho_{n-1} V_{n-1})/(\rho_n V_n + \rho_{n-1} V_{n-1})$$

where the suffix n refers to the nth layer. A log of reflection coefficients may thus be obtained and from it the ideal seismic reflection trace calculated.

Now, in essentially all exploration wells a Sonic Log is obtained. This gives the required velocity information. Density information is not so common but since density varies little in comparison with velocity and since, in any case, it normally increases and decreases in sympathy with velocity its absence is not often of great importance.

Once the synthetic seismogram has been obtained the next step is to compare it with those traces on the seismic section which were obtained at or very close to the well. However, before this can be done properly it is necessary to postulate a waveform for the seismic signal. This waveform is the impulse response of the total seismic system excluding the reflection coefficient log. That is, it is the series impulse response of the source, earth absorption, recording equipment and data processing. It is often about one tenth of a second in duration so that each individual reflection, both primary and multiple, lasts this long. Consequently, the 'reflection' seen on the seismic section does not result from a single interface but is affected by all the changes occurring over a finite depth interval of several tens of feet. The choice of waveform to be used is the most critical choice which has to be made when calculating a synthetic seismogram and the best method of making it is still a subject for research and development.

Once the waveform has been chosen the final synthetic seismogram is calculated and then plotted in the same manner as the field section. A detailed comparison of the two enables the interpreter

I. To identify those parts of the subsurface which generated the recorded reflections.

II. To identify any multiples on the field section, since they may be unambiguously identified on the synthetic seismogram with which the field section is being compared.

III. To help decide whether the data processing of the field section has been optimum, particularly with regard to deconvolution and band-pass filtering.

IV. To help decide whether the total recording and processing system has sufficient resolving power and noise rejection capability to solve the geological problem being asked. For instance, whether or not it will be possible to follow a particular thin reservoir sand.

V. Lastly, and most importantly, to predict how the field trace will alter should the subsurface alter in some particular way. For instance, in attempting to locate the feather edge of a pinch-out the sonic log may be modified so that the interval of interest is reduced in thickness and a new synthetic seismogram constructed. This is then compared with the initial synthetic and changes in the reflection character are noted and looked for on the field section. Of course, several synthetics should be constructed with progressively greater reduction in interval thickness so that the whole range of character change may be obtained as the interval varies from maximum thickness to zero. In such an exercise geological decisions will have to be made in order to specify in which order the different parts of the interval should be reduced or eliminated.

Initial efforts to use synthetic seismograms proved largely abortive and the technique fell into disrepute. However, with the advent of the Borehole Compensated Sonic Log, the dramatic increase in the signal to noise ratio on the seismic section and a better appreciation of the limitations of the technique – parallel layering, plane interfaces, zero offset, uniform absorption, etc – it is now undergoing a slow revival.

Should the interpreter decide to seek help from synthetic seismograms it is essential that he should confer closely with the geophysicist who produces them for, just like the seismic data processor, he makes many hidden judgements on the way to the final synthetic. These take place when choosing sea-bed or other near-surface reflection coefficients, when editing and calibrating the sonic log, when omitting/including surface multiples, when omitting/including earth absorption, etc etc, and their significance needs to be understood by the interpreter.

23 Zero-crossings for interpreters

23.1 Zero-crossings in general

The zero-crossings are the points where the time-series (i. e. seismic trace) or other series changes sign. Recording only zero-crossings is really the limiting case of the clipping of the data. The process is one employed in other fields when recording data of low amplitude, which is interrupted by occasional bursts of high amplitude noise. Amplitudes above a certain value are replaced in the series by that limiting value. It is a way of minimizing the effect of such noise in additive processes such as the seismic stacking process.

The 'hard clip' is a process which cuts all amplitudes down to unity, in effect to the sign or sign bit of each sample of the data. This leads into an elegant mathematical field, and it is surprising to see how much 'information', in the geological sense, remains when all the amplitude data have been discarded. It seems unlikely, however, that the procedures based on the 'hard clip' will ever yield more for the interpreter than processes making use of amplitudes. There are, however, 'hard clip' procedures which give almost as much for much less time and expenditure: and perhaps polarity-coincidence correlation is the best known of these.

There is one 'hard clip' process which has some interest for the interpreter. If one considers the final stack being made with hard clipped data, it will be seen that the amplitude of the output is an approximate measure of the signal to noise ratio. If there is no signal (in the sense of 1.1), the amplitudes are as likely to be $+1$ as -1, and the trace will vary irregularly about zero. At the peak of a powerful reflection, the ordinates will all be positive, and the output will be 48 for a 48-fold stack. Away from the peak there is more chance of noise overcoming the true value of the ordinate, and so reversing the polarity, so that the 'hard clip' stack output will drop. Multiples will have a lower amplitude than primaries in such an output, however high the multiple amplitudes may have been.

23.2 Zero-crossings of the T-series

It is sometimes a little helpful to display the zero-crossings on a final section instead of the regular peaks and troughs of a variable area display. This adds no new information, but it is sometimes an aid to its appreciation.

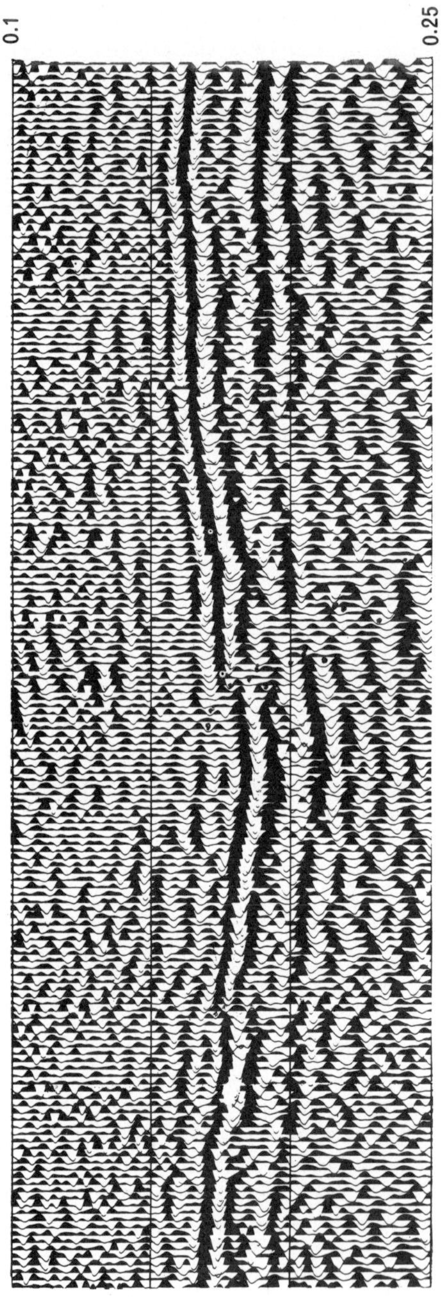
0.1
0.25

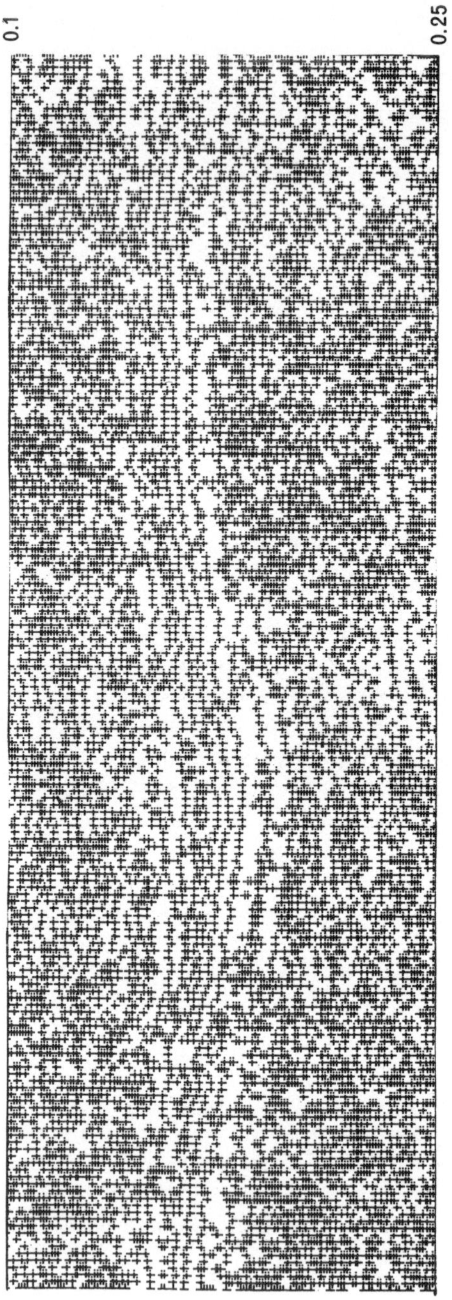
0.1
0.25

For example, the eye is so held by a powerful black reflection, that the phenomena displayed by much weaker reflections cannot be readily grasped. The zero-crossings will sometimes help this appreciation: and faults, which are inclined to the vertical, are emphasized by such zero-crossing displays. As with any procedure which effectively doubles the frequency of what is being examined, poor reflections are lost in noise, and only the best yield more to the eye of the interpreter. See fig. 56.

The doubling of frequency can be seen by considering a 100 Hz signal. It has 100 peaks per second, spaced at 10 ms. Its zero-crossings are spaced at 5 ms: so that if some signal or symbol is used to denote a zero-crossing one has, in effect, a 200 Hz signal.

This is a technique of very limited value.

23.3 Zero-crossings of the X-series

Several considerations lead one to expect zero-crossings in the X-direction in the neighbourhood of faults.

If there is no significant development of hyperbolas, the cross-section will have a 'butt-jointed' appearance at the fault. Many of the reflections, when traced laterally, pass into a reflection of opposite polarity across the fault. Many zero-crossings cluster along the fault plane.

Where hyperbolas are strongly developed, they change polarity at the seismic position of the fault. These polarity changes will also align themselves along the fault trace, and reinforce the indications of the first group. Whatever the precise configuration, the fault trace is indicated by cluster of polarity changes, or zero-crossings.

In a typical cross-section there are many factors producing zero-crossings in the X-direction. Dip, noise, or merely raggedness in the reflection alignment will all produce many zero-crossings, and swamp the fault indications. Consequently, a computer print-out of the zero-crossings along the X-axis is rarely of any use to the interpreter.

It is possible, however, to follow reflections manually, and to note the polarity change, or zero-crossing positions. This enables one, at times, to delineate the fault trace with precision. Fig. 43 is an example.

24 Averages, and other central measures

24.1 The nature of seismic noise

The interpreter is often disturbed by a situation in which he can find reflections over part of a spread on the raw data, but not at all on the final section.

Fig. 56. The upper section is a conventional one. It shows two small faults, one near the centre, the other near the left-hand end. The centre one has 5 ms throw, the left-hand one has 3 ms throw. A strong alignment marks the centre fault. This alignment is a little curious. It is not a fault plane reflection, which would be displaced rather far from the truncated ends of the beds. It is, rather, an alignment of the last peaks of each reflection along the line of truncation. No recognizable hyperbolas are developed here to confuse the picture.
The lower section plots the zero crossings of the same data. As often happens when the frequency is effectively doubled, many of the reflections disappear into noise, and only the most significant remain.
Viewed obliquely, the zero crossing section shows the same two faults. Perhaps the reverse fault character is more clearly shown, but there is otherwise little gained from the exercise.

There are many circumstances which can give rise to this kind of observation. The one dealt with here results from a failure to assess the nature of the seismic noise.

If noise is random in character, then the best estimate which can be made of the value of the reflection signal at a given time from an observation point is the mean of several observations. Should the nature of the noise be unknown, this may be a reasonable working assumption. It is the assumption which governs the traditional stacking or summing operations.

However, there are many cases, perhaps a majority of cases, where examination of the traces of a gather shows a quite different noise relationship. Some traces will show a reflection clearly visible. Others will show a large excursion due to noise – ground roll, a passing truck, electrical pick-up, or any of the other familiar noise sources. The noise may have many times the amplitude of the reflection – perhaps a hundred times.

Under these conditions, the use of a mean as the central measure is clearly unjustified: the signal-dominated traces are swamped by the noise-dominated traces. If there are more signal-dominated traces than noise-dominated traces, then the median is a more representative central measure: it is an analytical method for recognizing and discarding the high amplitude noise traces. It has the further advantage that if these noise relationships change along the traverse, and the noise becomes random, then the median and the mean have very similar values, especially if the data is recorded with a high multiplicity.

Under some special conditions, other central measures may be preferable. For example, if the noise traces are few, then a central quartile might be the better selection of data values, and its mean the best estimate of the value of the ordinate. Perhaps, under some noise condtions the mode, the most commonly occuring value is the best central measure. Some experiments have been made with the median, but not, as far as the writer can discover, with other central measures.

25 Contouring of structure maps and three-dimensional presentation

25.1 Making use of the known shapes of folds

Contouring has been described in a preliminary way in chapter 14 of Level 2. The return to the subject is an opportunity to stress the essential contribution of the interpreter, and to separate it from the large part of the process which can be mechanized.

The interpreter's decisions are those of joining up corresponding structural elements on successive lines of traverse. This includes faulting: no analytical procedures are known which will join one fault indication to another. With very closely-spaced traverses it is possible that pattern recognition would help: but mostly it is a geological decision, based on a knowledge of fault patterns and the geometry of faulting.

Similarly, in the case of fold axes, analytical methods can be used with closely-spaced traverses: but in general the decisions to join up fold axes is taken from a knowledge of strike direction or an awareness of the tectonic style of the region.

An application of the ideas of 13.6 (Level 2) would lead to an approach which could be summarized in the following steps

I. Plot the data for the horizon to be contoured.
II. Draw in faults and fold axes, and lines of inflexion.
III. Analyse each block of data on the line of traverse (i. e. as bounded by the lines of II.), into a principal sine waveform and its harmonics.
IV. Where the same block is cut on two successive parallel traverses, interpolate by assuming linear changes in wavelength and amplitude of the principal sine wave and of its harmonics.
V. Where the same block is cut by two traverses at right angles or some other intersection angle, project the wavelengths of principal sine wave and harmonics on to the full dip line, and interpolate as before.

This approach may give machine contours which are more acceptable than those based on current programmes.

25.2 Stereo presentation

Standard methods exist for drawing mechanically the two versions of a set of contours, which, when combined stereoscopically, produce a three-dimensional image.

Most interpreters see their data in three dimensions in any case, and the stereo view adds but little to their visualization – which cannot be better than the interpolation procedure used. In some other geophysical work, principally in aeromagnetics, it has been noted that linear features of small amplitude, superimposed on a rugged background, can be recognized stereo-scopically although they cannot be detected on a flat contour presentation. No such advantage has been demonstrated for seismic data – perhaps because of imperfect contouring.

25.3 An alternative approach to seismic survey

An alternative approach to seismic survey, of great technical interest, is the holographic method. In some forms that the experimental work has taken, the data is recorded three-dimensionally.

It is difficult to forecast the future developments of this subject, though it seems unlikely to take on a dominant role in seismic exploration.

The subject is a specialized one and reference should be made to the original papers. (See Bibliography under Level 3).

26 Branched VT curves

26.1 Branched reflections and VT curves

A passing reference was made to branched T-X curves in Level 3, 19.7.

Branched velocity curves result inevitably from branched reflections. Both are an expression of the fact that the locus of normal reflection points from a single observation point has itself branched.

Such branching is a feature of seismic work in areas of complex structure. At greater depths, structures of much gentler character give rise to branching. At the greatest depths at which seismic work is carried out, branching is the rule, and a single VT curve (and a single reflection) is the exception.

At the present time, this branching is largely ignored. Velocities used for stacking and for interval velocity studies are often a slurred version of the results on the several branches, and this must be erroneous. It seems likely that the path of progress will be to separate the branches, and perhaps to process separately to benefit the several reflection branches.

27 Some limits to automation

27.1 A discussion of the essential contribution of interpreter and machine

It is implied throughout this work on interpretation that there is a clearly defined field for the interpreter. As a final chapter this thesis will be examined more closely.

Current developments in seismic are all in the direction of automation, of computer usage. This leads to savings in time, money and effort: clearly this must advance as far and as fast as the technology will allow. So the definition of the role of the interpreter really turns on whether automation will proceed quickly and smoothly to the end, and eliminate the interpreter. Or is there some other limitation which will preserve a field for the interpreter, at least for a while.

There is such a limitation. There are interpreter functions which must exercise geological judgements. They are:

Drawing continuity and making correlations.
Assessing the significance of velocities.
Drawing fold axes and faults.

In carrying out these tasks the interpreter is exercising geological judgements. Decisions are taken which are not necessarily self-evident from the data. It can be argued that the basis for these decisions will be codified, and that the decision-making can then be programmed. This may well be done if interest in seismic interpretation survives so long.

On the other hand, all the other tasks, which do not require geological judgements, are more readily expressed in that logical form which is the starting point of the programmer's work. Much remains to be done, but the rate of progress is considerable and can be assessed from the contents of specialist journals. With this progress, more and more of the 'other' tasks can be programmed: and there is need to accelerate these mechanical aids to the interpreter, and to reduce his load of mechanical work as far as possible.

Bibliography

Level 1

BADGLEY, P. C. (1973): Structural methods for the exploration geologist. – Harpers.

BARANOV, V. & PICOU, C. H. (1964): Energy and vector record-sections. – Geophysics, XXIX: 17 – 37.

BOIS, P. & LAPORTE, M. (1970): Pointe automatique. – Geophys. Prospect. XVIII: 489 – 504.

BOULWARE, R. A. (1967): Seismic exploration for stratigraphic traps in Canada. – Proc. 7th World Congress, p 471.

BRAITHWAITE, C. J. R. (1973): Reefs: just a problem of semantics. – Bull. AAPG, LVII: 1100.

BUCHHOLTZ, H. (1972): A note on signal distortion due to dynamic (NMO) corrections. – Geophys. Prospect. XX: 395 – 402.

BURCHFIELD, B. C. & DAVIS, G. A. (1973): Possible igneous analog of salt dome tectonics, Clark Mountains, South Eastern California. – Bull, AAPG, LVII: 993.

BUSK, H. G. (1929): Earth flexures. – Cambridge University Press.

CLAERBOUT, J. F. & DOHERTY, S. M. (1972): Downward continuation of move-out corrected seismograms. – Geophysics, XXXVII: 741 – 768.

CLOWES, R. M., KANASEWICH, E. R. & CUMMING, G. L. (1968): Deep crustal seismic reflections at near vertical incidence. – Geophysics, XXXIII: 441 – 451.

COHEE, G. V. (1972): Reports published by International Subcommission on Stratigraphic Classification. (A Review). – Bull. AAPG, LVI: 1135.

CUMMINGS, E. R. (1932): Reef or bioherms. – Geol. Soc. Amer. Bull., XLIII: 331.

DAVIS, T. L. (1972): Velocity-density variations around Keg River Reefs, Alberta. – Journal of the Canadian Soc. of Economic Geophysicists, I: No. 1.

DAVIS, T. L. (1972): Velocity variations around Leduc Reefs Alberta. – Geophysics, XXXVII: 584 – 604.

Delta Symposium, (1971): A series of papers. – Bull. AAPG, LV: 1135 et seq.

Deltas (1966): – Houston Geological Soc.

DMITRIYEV, V. A. (April 1971): On the spectral composition of underwater seismic noise in coastal regions of the sea. – Bull. Izvestia Academy of Sciences USSR. Physics of the Solid Earth.

DOHR, G. (1971): First experience and results with a new seismic interpretation method. – Geophys. Prospect. XIX: 371 – 387.

EVENDEN B. S., STONE, D. R. & ANSTEY, N. A. (1970): Seismic prospecting instruments. – Vol. I: ANSTEY, N. A. Signal characteristics and instrument specifications. – Geopublication Associates. Stuttgart.

FALLAW, W. C. (1973): Grabens on anticlines in Gulf Coastal Plain, and thinning of sedimentary section in downthrown fault block. – Bull. AAPG, LVII: 198.

FRANKLIN, J. N. (1968): Matrix theory. – Prentice Hall.

GAROTTA, A. (1971): Selection of seismic picking based upon the dip, move-out and amplitude of each event. – Geophys. Prospect. XIX: 357 – 370.

GOFORTH, T. T., DOUZE, E. J. & SORRELLS, G. G. (1972): Seismic noise measurements in a geothermal area. – Geophys. Prospect. XX: 76 – 82.

GOLD, B. & RADER, C. M. (1969): – Digital processing of signals. p 20 et seq., p 98 et seq. – McGraw Hill.

GUZMAN, E. J. (1967): Reef type stratigraphic traps in Mexico. – Proc. 7th World Petroleum Congress.

HAMILTON, E. L. (1970): Reflection coefficients and bottom losses at normal incidence computed from Pacific sediment properties. – Geophysics, XXXV: 995 – 1004.

HARMS, J. C. & TACKENBERG, P. (1972): Seismic signatures of sedimentation models. – Geophysics, XXXVII: 45 – 58.

HILTERMAN, F. J. (1970): Three-dimensional seismic modelling. – Geophysics, XXXV: 1020 – 1037.

HINTZE, W. H. (1971): Depiction of faults on stratigraphic isopach maps. – Bull. AAPG, LV: 871.

HOME, G. (1948): Roman London. – Ernest Benn.

HRISKEVICH, M. E. (1970): Middle Devonian Reef Production Rainbow Area, Alta. – Bull AAPG, LIV: 2260.
HUBBARD, M. K. (1973): Structural geology. – Hafner.
IYER, H. M. (1973): Seismic noise in geothermal areas. Abstract only. – Geophysics, XXXVIII: 177.
IYER, H. M. & HITCHCOCK, T. (1974): Seismic noise measurements in Yellowstone National Park. – Geophysics, XXXIX: 389 – 400.
International Subcommission on Stratigraphic Classification Reports (1961 – 1971):

ISSC Report 1 Principles of stratigraphic classification and terminology.
Pt. 25, 21st Int. Geol. Cong. Proc. 1961.
ISSC Report 3 Preliminary report on lithostratigraphic units.
24th Int. Geol. Cong. 1971.
ISSC Report 5 Preliminary report on biostratigraphic units.
24th Int. Geol. Cong. 1971.
ISSC Report 6 Preliminary report on chronostratigraphic units.
24th Int. Geol. Cong. 1971.

JANKOWSKY, W. (1970): Empirical investigation of some factors affecting elastic wave velocities in carbonate rocks. – Geophys. Prospect. XVIII: 103 – 118.
JOHNSON, R. C. & THOMSON, J. T. (1973): Streamer tow-noise: Post acceleration-cancelling hydrophone era (Abstract only). – Geophysics, XXXVIII: 178.
JOLLY, R. N. (1973): Random noise from seismic sources (Abstract only). – Geophysics, XXXVIII: 178.
KRUMBEIN, W. C. & SLOSS, L. L. (1963): Stratigraphy and sedimentation. esp. p 575 et seq. – W. H. Freeman & Co.
LEE, Y. W. (1960): Statistical Theory of Communication. – John Wiley 5 Sons Inc.
LEVIN, F. K. & SHAH, P. M. (1973): Gross properties of time-distance curves. – Geophysics, XXXVIII: 643.
LOHMANN, H. H. (1972): Salt dissolution in subsurface of British North Sea as interpreted from seismograms. – Bull. AAPG, LVI: 472.
MARR, J. D. (1971): Seismic stratigraphic exploration Pt. III: 676 – 689.
MASCLE, G. MASCLE, J. (1972): Aspects of some evaporitic structures in Western Mediterranean Sea. – Bull. AAPG, LVI: 2260.
MAURIN, A. F. & RAASCH, G. O. (1972): Early-Frasnian stratigraphy, Kakwa-Cecilia Lakes British Columbia – Canada. – Notes et Memoires No. 10 publ. Comp. Francaise des Petroles, esp. p 35 et seq.
MCLAUGHLIN, D. H. (1972): Evaporite deposits of Bogota area, Cordillera Oriental, Columbia. – Bull. AAPG, LVI: 2240.
MEEK, J. W. (1973): Three-dimensional model – Gulf of Mexico. – Geophysics, XXXVIII: 295 – 300.
MYERS, J. B. (1972): Permian patch reefs in Finlay Mountains, West Texas. – Bull. AAPG, LVI: 494.
O'DOHERTY, R. F. & ANSTEY, N. A. (1971): Reflections on amplitudes. – Geophys. Prospect. XIX: 430 – 458.
OXLEY, M. L. & HERLIHY, D. E. (1972): The Bryan Field – a sedimentary anticline. – Geophysics, XXXV: 59 – 67.
PATURET, D. (1971): Different methods of time-depth conversion with and without migration. – Geophys. Prospect. XIX: 27 – 41.
PAULSON, K. V. & MERDLER, S. C. (1968): Automatic seismic reflection picking. Geophysics, XXXIII: 431 – 440.
PRESS, F. (1966): Seismic velocities in Handbook of Physical Constants. – Geol. Soc. America, Memoir 97: 195.
RICHARDS, P. G. (1971): Elastic wave solutions in stratified media. – Geophysics, XXXVI: 798 – 809.
ROBINSON, E. A. (1967): Statistical communication and detection. – Charles Griffin & Co. Ltd.
SCHNEIDER, E. D. & JOHNSON, G. L. (1970): Deep-ocean diapyr occurrences. – Bull. AAPG, LIV: 2151.
SCHWARTZ, R. J. & FRIEDLAND, B. (1965): Linear systems. – McGraw Hill.
SMITH, D. A. & REEVE, F. A. E. (1970): Salt piercement in shallow Gulf Coast Structures. – Bull. AAPG, LIV: 1271.
STEWART, J. L. (1960): Fundamentals of signal theory. – McGraw Hill.
TANER, M. T. & KOEHLER, F. (1969): Velocity spectra – digital computer derivation and applications of velocity functions. – Geophysics, XXXIV: 859 – 881.
TROREY, A. W. (1970): A simple theory for seismic diffractions. – Geophysics, XXXV: 762 – 784.
TRUSHEIM, F. (1960): Mechanism of salt migration in Northern Germany. – Bull. AAPG, XLIV: 1519.
TUCKER, P. & YORSTON, H. (1973): Pitfalls in seismic interpretation. – SEG Monograph No. 2. 1973.
WALKER, J. R. & ENSMINGER, H. R. (1970): Effect of diapirism on sedimentation in Gulf of Mexico. – Bull. AAPG, LIV: 2058.
WATKINS, J. H., WATERS, L. A. & GODSON, R. H. (1972): Dependence of in-situ compressional wave velocity on porosity in unsaturated rocks. – Geophysics, XXXVII: 29 – 35.
WIDESS, M. B. (1943): Multiple branches in seismic reflection-time surfaces. – Geophysics, VIII: 93 – 104.

WILLIAMS, A. B. (1973): Computer aided interpretation of gravity anomalies with two-dimensional models derived from seismic profiles (Abstract only). – Geophysics, XXXVIII: 195.

Level 2

ALLENBY, R. J. (1962): The importance of reflected refractions in seismic interpreting. – Geophysics, XXVII: 966 – 980.

BADING, R., ECHTERHOFF, J., KREY, T. & MARSCHALL, R. (1974): New possibilities for reflection seismics by undershooting bodies with complicated tectonics. – Geophys. Prospect. XXII: 1 – 21.

BENTLEY, C. R. (1972): Seismic wave velocities in anisotropic ice: a comparison of measured and calculated values in and around the deep drill hole at Byrd Station, Antarctica. – J. Geophys. Research, LXXVII: 4406.

BORTFIELD, R. & HÜRTGEN, H. (1960): On the identification and construction of reflected refractions. – Geophys. Prospect. VIII: 12 – 24.

FRENCH, W. S. (1974): Two-dimensional and three-dimensional migration of model-experiment reflection profiles. – Geophysics, XXXIX: 265 – 277.

GARDNER, G. H. F., FRENCH, W. S. & MATZUK, T. (1973): Elements of migration and velocity analysis. – This paper was presented at the 43rd meeting of SEG, Mexico City, 1973.

HAGEDOORN, J. G. (1954): A process of seismic reflection interpretation. – Geophys. Prospect. II: 85 – 127.

HAMILTON, E. L., BUCKER, H. P., KEIR, D. L. & WHITNEY, J. A. (1970): Velocities of compressional and shear waves in marine sediments determined in situ from a research submarine. – J. Geophys. Research, LXXV: 4039.

HILDEBRAND, F. B. (1956): Introduction to Numerical Analysis (esp. chapters 1 – 4).

HOSKING, A. R. (1966): A regional investigation to determine time-depth and velocity-depth relationships applicable to Central Australia. – Australian Petroleum Expl. Assoc. J.: 50.

JUNKINS, J. L., MILLER, G. W. & JANCAITIS, J. R. (1973): A weighting function approach to modelling of irregular surfaces. – J. Geophys. Research, LXXVIII: 1794.

KELLY, K. R., ALFORD, R. M., TREITEL, S. & WARD, R. W. (1974): Application of finite-difference methods to exploration seismology. – Paper presented to the Royal Irish Academy Conference on Numerical Analysis.

LASKI, J. D. (1970): Simultaneous estimation of parameters of reflection events (depth, dip, velocity) and relative static corrections. – Geophys. Prospect. XVIII: 269 – 276.

MARSHALL, R. (1972): Parameter optimization of velocity depth functions of given form by use of root-mean-square velocities. – Geophys. Prospect. XX: 700 – 711.

MCDONAL, F. J., ANGONA, F. A., MILLS, R. L., SANGBUSH, R. L., VAN NOSTRAND, R. G. & WHITE, J. E. (1958): Attenuation of shear and compressional waves in Pierre Shale. – Geophysics, XXIII: 421 – 439.

NEWMAN, P. (1973): Divergence effects in a layered earth. – Geophysics, XXXVIII: 481 – 488.

O'DOHERTY, R. F. & ANSTEY, N. A. (1971): Reflections on amplitudes. – Geophys. Prospect. XIX: 430 – 458.

PILBEAM, C. C. & VAISNYS, J. R. (1973): Acoustic velocities and energy losses in granular aggregates. – J. Geophys. Research, LXXVIII: 810.

ROBINSON, J. C. (1970): An investigation of the relative accuracy of the most common normal-moveout expressions in velocity analyses. – Geophys. Prospect. XVIII: 352 – 363.

SHERIFF, R. E. (1973): Factors affecting amplitudes – a review of physical principles. In Lithology and direct detection of hydrocarbons using geophysical methods. – Symposium sponsored by The Geophysical Society of Houston.

SHERWOOD, J. W. C. & POE, P. H. (1972): Continuous velocity estimation and seismic wavelet processing. – Geophysics, XXXVII: 769 – 787.

SHETRONE, H. A. (1972): Computer mapping of seismic and velocity data in Australia. – Geophysics, XXXVII: 313 – 324.

STABLER, C. L. (1968): Simplified Fourier Analysis of fold shapes. – Tectonophysics, VI, No. IV: 343 – 350.

STOLL, R. D., EWING, J. & BRYAN, G. M. (1971): Anomalous wave velocities in sediments containing gas hydrates. – J. Geophys. Research, LXXVI: 2090.

TANER, M. T., COOK, E. E. & NEIDELL, N. S. (1970): Limitations of the reflection seismic method; lessons from computer simulations. – Geophysics, XXXV: 551 – 573.

TEGLAND, E. R. (1973): Computer assisted interpretation of seismic structure and velocity data (Abstract only). – Geophysics, XXXVIII: 192.

TERNUS, R. A. (1973): An adaptive algorithm for the determination of velocity and structural models from seismic data (Abstract only). – Geophysics, XXXVIII: 192.

WREN, A. E. (1973): Contouring and the contour map: a new perspective (Abstract only). – Geophysics, XXXVIII: 196.

WREN, A. E. (1973): Trend surface analysis – a review. – J. of the Canadian Society of Economic Geophysics, IX, No. I: 39.

Level 3

AL-CHALABI, M. (1973): Series approximation in velocity and travel-time computations. – Geophys. Prospect. XXI: 783 – 795.

BARANOV, V. & KUNETZ, G. (1960): Film synthetique avec reflexions multiples theorie et calcul pratique. – Geophys. Prospect. VIII: 315 – 325.

BECKER, A. A. (1974): Seismic expression of Pre-Tertiary structures in East and South Texas. – This paper was read at the SEG meeting in Dallas, Texas, November 1974.

BOGERT, B. P., HEALY, M. J. R. & TUKEY, J. W. (1963): A quefrency analysis of time series for echoes: Cepstrum, Pseudo-autocovariance, Cross-cepstrum and Saphe Cracking. – In Proceedings of the Symposium on Time Series Analysis, New York, p 209.

CAUGHLIN, W. G., LUCIA, F. J. & MCIVER, N. L. (1974): The detection and development of Silurian Reefs in Northern Michigan. – This paper was read at the SEG meeting in Dallas, Texas, November 1974.

CODET, P., DAMOTTE, B., ERB, E., LAVERGNE, M. & LAYOTTE, P. C. (1971): Sismique minière à haute frequence. – Geophys. Prospect. XIX: 776 – 790.

FAUST, L. Y. (1951): Seismic velocity as a function of depth and geologic time. – Geophysics, XVI: 192 – 206.

FONTANEL, A. & GRAU, G. (1969): Application of impulse seismic holography. – Paper read to Society of Exploration Geophysicists. – Institut Francais du Petrole reprint.

GASSMAN, F. (1951): Elastic waves through a packing of spheres. – Geophysics, XVI: 673 – 685.

HAGEDOORN, J. G. (1962): In pursuit of the errant seismic pulse. – Geophys. Prospect. X: 148 – 165.

HOFER, H. & VARGA, W. (1972): Seismogeologic experience in the the Beaufort Sea. – Geophysics, XXXVII: 605 – 619.

HORNABROOK, J. T. (1974): Seismic re-interpretation clarifies. – Petroleum International, XIV: 45 – 53.

KEMERAIT, R. C. & CHILDERS, D. G. (1972): Signal detection and extraction by cepstrum techniques. – IEEE Transactions on Information Theory, Col. 1T-18, p 745.

KOEFOED, O. (1960): Measurement of amplitudes of reflected seismic waves. – Geophys. Prospect. VIII: 25 – 26.

KOEFOED, O. (1962): Reflection and transmission coefficients for plane longitudinal incident waves. – Geophys. Prospect. X: 304 – 351.

KOEFOED, O., OOSTERVELD, M. M. & ALONS, I. J. G. (1963): A laboratory investigation into the elastic properties of limestones. – Geophys. Prospect. XI: 300 – 312.

LEHNER, P. (1969): Salt tectonics and Pleistocene stratigraphy on continental slope of northern Gulf of Mexico. – Bull. AAPG. LIII: 2431 – 2479.

LERWILL, W. E. (1969): Holography at seismic frequency. – European Association of Exploration Geophysicists, Venice.

MICHON, D., WLODARCZAK, R. & MERLAND, J. (1971): A new method of cancelling multiple reflections: "Souston". – Geophys. Prospect. XIX: 616 – 625.

MORGAN, N. A. (1907): Wavelet maps: a new analysis tool for reflection seismograms. – Geophysics, XXXV: 447 – 460.

MUELLER, R. K. (1971): Acoustic Holography. – Proc. of the IEEE, p 1319 – 1335.

MUIR, F. & HALES, F. W. (1955): A rational approach to the design of electrical filters and of shot-hole and geophone patterns in seismic reflection prospecting. – Geophys. Prospect. III: 350 – 358.

O'BRIEN, P. N. S. (1969): Some experiments concerning the primary seismic pulse. – Geophys. Prospect. XII: 511 – 547.

PETERSON, R. A. Seismography 1970: The writing of the earth waves. – Symposium, Society of Exploration Geophysicists on Seismic Waves from Vibrational Sources. Pacific Coast Section, Los Angeles. 1969.

PETERSON, R. A. (1974): Through the Kaleidoscope. – Bendix reprint.

PETERSON, R. A. et al. (1955): Synthesis of seismograms from well log data. – Geophysics XX: 516 – 538.

RICKER, N. (1951): The form and laws of propagation of seismic wavelets. – Proc. of the 3rd World Petroleum Congress. Section I, 514.

SCHNEIDER, W. A., PRINCE, E. K. & GIBS, B. F. (1965): A new data processing technique for multiple attenuation exploiting differential normal move-out. – Geophysics. XXX: 348.

SHERIFF, R. E. (1973): Encyclopedic Dictionary of Exploration Geophysics. – A bilingual (Spanish-English) version of this was produced by Seismograph Service Corporation and distributed at the 1973 meeting of SEG in Mexico City.

TANER, M. T., KOEHLER, F. & ALBILALI, K. A. (1974): Estimation and correction of near surface time anomalies. – Geophysics, XXXIX: 441 – 463.

Index

GEOEXPLORATION MONOGRAPHS

Several excellent text books exist which cover the subject of Applied Geophysics but the degree of specialisation and sophistication of present geophysical techniques inevitably means that they have insufficient detail to be used other than as a general introduction.

GA intends to provide a series of monographs with sufficient detail to be of use to the practising geophysicist and with sufficient discussion of the scientific fundamentals to be suitable for the student and the research worker.

Essentially, the monographs will be critical reviews of the aims, methods and status of the various techniques covering the whole range of instrumentation, operation and interpretation. They will each be written by a practising geophysicist of considerable experience.

The monographs appeal to advanced students in geoexploration at Universities and Technical High Schools as well as to geophysicists and others engaged in geophysical exploration.

Geopublication Associates

N. A. Anstey, London — H. Braekken, Trondheim — G. Grau, Paris
O. Koefoed, Delft — G. Kunetz, Paris — H. Menzel, Hamburg
C. Morelli, Trieste — R. G. van Nostrand, Alexandria/Virginia
P. N. S. O'Brien, London — D. S. Parasnis, Boliden
O. Rosenbach, Clausthal — S. Saxov, Aarhus

GEBRÜDER BORNTRAEGER · BERLIN · STUTTGART

GEOEXPLORATION MONOGRAPHS

Principles of Direct Current Resistivity Prospecting Series 1, No. 1

by GEZA KUNETZ (Editors: H. BRAEKKEN — R. VAN NOSTRAND). 8°, 58 figures. XII, 106 pages. 1966. Cloth bound DM 20,—.

Contents: Introduction – Equipotential Maps – Resistivity Profile and Maps – Electrical Soundings – Determination of the Potential Distribution in a layered Medium – Practical Calculation of Apparent Resistivities.

The Application of the Kernel Function in Interpreting Geoelectrical Resistivity Measurements Series 1, No. 2

by O. KOEFOED, Professor of exploration geophysics at the Technological University at Delft (Editors: G. KUNETZ—S. SAXOV). 8°. XII, 111 pages, 47 figures, 9 plates. 1968. Cloth bound DM 36,—.

Contents: The fundamental relation between the electrical potential at the surface and the distribution of resistivity layers in the subsurface – The apparent resistivity and its relation to the kernel function – Determination of the kernel function by the decomposition method – Derivation of the partial kernel functions – Partial apparent resitivity functions for the WENNER configuration – Practical applications of determination of the raised kernel function – Introduction on determinants – The relation between the kernel function and the dephts and resistivity contrasts of the boundary planes – First method of determining the layer distribution from the raised kernel function – Second method of determining the layer distribution from the raised kernel function – Some special procedures – The inverse problem: prediction of the detectability of subsurface layers – Appendix: The mathematical expressions for the partial apparent resistivity functions and the partial kernel functions.

Seismic Prospecting Instruments Series 1, No. 3

by B. S. EVENDEN, D. R. STONE and N. A. ANSTEY (Editors: G. GRAU— P. N. S. O'BRIEN).

Vol. 1: Signal Characteristics and Instrument Specifications

by N. A. ANSTEY. 8°. X, 156 pages, 113 figures. 1970. Cloth bound DM 41,—

Contents: Elements of the seismic prospecting method – The manipulation of waveforms: Basic waveform concepts and definitions – The specification of waveforms – Linear operators and convolution – The seismic reflection technique: The seismic signal – Noise and surface waves – The field arrangement – The seismic reflection record – The complete seismic chain – Desirable instrument characteristics: The easy ones – The problem of non-linearity – The problem of system noise – The final compromise specification – Appendix – Bibliography – Index.

GEOEXPLORATION MONOGRAPHS

Vol. 2: Instrument Performance and Testing

by B. S. EVENDEN and D. R. STONE. 8°. XII, 195 pages, 159 figures. 1971. Cloth bound DM 47,—.

Contents: Seismic detectors – Analogue amplifiers and filters – Analogue recording on magnetic tape – Digital recording on magnetic tape – The visual recorder – Auxiliary equipment – Bibliography – Appendix: Binary and BCD conversion table – Index.

Geothermics with Special Reference to Application Series 1, No. 4

by O. KAPPELMEYER and R. HAENEL (Editors: O. ROSENBACH — C. MORELLI). 8°. X, 238 pages, 123 figures. 1974. Cloth bound DM 76,—.

Contents: Fundamental concept of heat transfer – Thermal conditions of the earth's interior – Thermal conditions of the earth's crust – Practical application of geothermics – Geothermal investigations in deep wells – Tables – Bibliography – Subject Index.

Principles of Magneto-Telluric Prospecting Series 1, No. 5

by G. PORSTENDORFER (Editors: G. KUNETZ — D. S. PARASNIS). 8°. VIII, 118 pages, 71 figures and 3 tables, 1974. Cloth bound DM 64,—.

Contents: Introduction – The natural electromagnetic field of the earth – Basic theoretical problems – Processing of field records – Digital processing of field records – Instruments and methods for field recording – Geological examples – Concluding remarks on the accuracy of magneto-telluric investigations and interpretations – Literature – Index.

Potential Fields and their Transformations in Applied Geophysics Series 1, No. 6

by WLADIMIR BARANOV (Editors: R. G. VAN NOSTRAND – S. SAVOX). 8°. XVI, 121 pages with 32 figures and 18 tables. 1975. Cloth bound DM 64,—.

Contents: I. Physics of fields and potentials: Fields and potentials – Integral formulae and harmonic functions – Meaning of equation for harmonic functions in applied geophysics – Magnetism in applied geophysics. – II. Functional transformation in applied geophysics: Solution of Laplace's equation – Functions with limited spectra – Filtering and functional transformations – Continuation of harmonic functions – Reduction of magnetic anomalies to the pole – Derivatives of harmonic functions – Establishing a regular grid – References – Index.